T0274423

THE CANCER
FACTORY

THE CANCER FACTORY

INDUSTRIAL CHEMICALS,
CORPORATE DECEPTION, AND
THE HIDDEN DEATHS
OF AMERICAN
WORKERS

JIM MORRIS

BEACON PRESS
BOSTON

Beacon Press
Boston, Massachusetts
www.beacon.org

Beacon Press books
are published under the auspices of
the Unitarian Universalist Association of Congregations.

27 26 25 24   8 7 6 5 4 3 2

This book is printed on acid-free paper that meets the uncoated paper
ANSI/NISO specifications for permanence as revised in 1992.

Text design and composition by Kim Arney

*Library of Congress Cataloging-in-Publication
Data is available for this title.*
ISBN: 978-0-8070-5914-2;
e-book: 978-0-8070-5915-9;
audiobook: 978-0-8070-1464-6

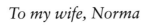

To my wife, Norma

CONTENTS

INTRODUCTION

A ROUND FOUR O'CLOCK in the afternoon on August 4, 1992, during a bath-room break from moving furniture, Rod Halford began urinating blood. Not a little blood; sheets of it. It was "such a startling effect," he would later testify in a deposition, that "I jumped back and [the blood] went all over." Halford, then living in Youngstown, New York, near Lake Ontario, had the presence of mind to hold his urine stream and ask his wife, Nellie, to bring him a jar so he could capture a sample. She brought him a baby bottle. Halford called his family doctor, who told him to dump the bottle's contents and report to Mount St. Mary's Hospital in Lewiston, seven miles away, the next morning. There Halford gave another urine sample, but a proper analysis wasn't possible—still too much blood. X-rays were taken two days later, and a urologist at Sisters of Charity Hospital in Buffalo gave Halford the results on August 10: he had a malignant tumor on his bladder the size of his pinky fingernail. The tumor was removed without general anesthesia. Halford had vetoed the idea of being put under and staying overnight in the hospital because, he explained in his deposition, Nellie couldn't drive his pickup truck, which had a stick shift, and "I just don't relish hospitals anyway." There was, Halford admitted, some discomfort associated with the procedure.

At the time of his diagnosis, Halford was the eighteenth employee of the Goodyear Tire and Rubber Company's chemical plant in Niagara Falls, New York, to develop bladder cancer. Halford had started at the plant as a chemical operator in 1956, become an electrician in 1970, and retired as a janitor at fifty-five in 1991. His tumor was, to some extent, foreseeable: by 1981, when he was president of Oil, Chemical and Atomic Workers Local 8–277, he knew of four coworkers with the disease. By 1988 he knew of eight. Federal health investigators, invited into the plant by the union, found fifteen cases of bladder cancer. At this writing, the unofficial tally stands at seventy-eight, though that may be an undercount given the challenges of tracking retirees. Some of the victims are dead. Others live with the anxiety of knowing their malignancies, contained by surgery and chemotherapy, could resurface at any time.

There's little, if any, doubt about what caused the Goodyear cancer cluster. From 1957 onward, Rod Halford and his coworkers in Niagara Falls inhaled—and, more important, absorbed—a chemical called ortho-toluidine, known internally as Dominic. The chemical was used to make an anti-cracking agent for tires. Halford would come home from work stinking of it. It made his wife gag when she washed his clothes. It seeped through his pores while he slept, leaving a brown stain on the sheets. Documents show its primary manufacturer, E.I. du Pont de Nemours and Company, knew by the 1950s that the yellowish liquid caused bladder cancer in laboratory animals, and was protecting its own production workers in New Jersey against exposure. But DuPont either didn't adequately warn its customer, Goodyear, about the risks, or Goodyear didn't listen. Maybe both companies were at fault; that discussion may never be settled.

Not large by the standards of its neighbors, the Goodyear plant had a history with chemicals that made it a pernicious, two-headed beast: eleven years before Goodyear started buying ortho-toluidine from Du-Pont and a few other manufacturers, it began using a chemical called vinyl chloride to make polyvinyl chloride (PVC), a resin fabricated elsewhere into vinyl garden hoses, shower curtains, siding, and other products. Like ortho-toluidine, vinyl chloride is a potent carcinogen, though it targets the liver instead of the bladder. By the time Goodyear closed the PVC side of the plant in 1996, eleven fatal cases of liver cancer had been recorded among its workforce—about four times what would be expected under normal circumstances.

Few outbreaks of occupational illness in the US have been documented as thoroughly as the one at Goodyear's Niagara Falls factory. Few offer a more striking example of how blue-collar workers were exploited after World War II. Generations of them—starting with the tide of veterans fresh from the battlefields of Europe and the Pacific, eager to ride the surging industrial economy—were misled about the safety of the chemicals they made and used. Men who had avoided death or serious injury in the Hürtgen Forest or the jungles of Guadalcanal encountered in workplaces like Goodyear an adversary they couldn't see, one that weakened untold numbers of them over time and lopped years off their lives. This was the bargain they unknowingly struck in exchange for company picnics, Thanksgiving turkey giveaways, and paychecks big enough, with overtime, to allow the purchase of fishing boats and weekend cabins. Many stayed in the same factory for years and believed what their employers told them: "Substance X won't hurt you." Or, in extreme cases: "You could eat Substance Y for breakfast and be fine." Employers and their trade associations, we now know, understood the enormity of such falsehoods.

The federal government gave this generation of workers almost no help until the late 1960s, by which time some 14,500 of them were dying of traumatic injury each year and tens of thousands more were being hurt or sickened. The need for a national worker-safety law had been recognized at least as far back as the Lyndon B. Johnson administration, but Congress didn't pass one until it was nudged by Richard Nixon early in his first term. The law created the Occupational Safety and Health Administration (OSHA), a regulatory agency that was starved of resources and vilified from the start; overwrought critics likened it to the Gestapo and the Spanish Inquisition. Apart from a burst of productivity in the late 1970s, the agency has largely faltered in its attempts to control dangerous substances in the workplace. It has set exposure limits for fewer than 1 percent of the chemicals ever made or used in the United States, and most of these limits aren't protective. Cancer and other work-related illnesses still take as many as 120,000 lives annually in this country—more than influenza, pneumonia, and suicide combined in 2019, the last full year before the COVID-19 pandemic struck.

Another two hundred thousand to four hundred thousand people get sick each year from on-the-job exposures. In truth, workers of all races and ethnicities in the United States are victims of legal segregation: they can be exposed to carcinogens at levels ten to one thousand times higher than what would be allowed just beyond the plant fence.

Niagara Falls, whose population has fallen by more than half since 1960, bet on heavy industry and lost. After World War II, chemical manufacturers flocked to the city and its cheap hydroelectric power. Jobs were so easy to find that, it was said, a worker laid off in the morning could be employed again by afternoon. Then companies began to flee to other states, other countries. With them went the workers and their families. An urban-renewal project failed, a toxic-waste dump came to personify the city, and Niagara Falls devolved into a den of poverty and crime. It is attempting yet another comeback, having attracted some modest commercial development downtown. Under no plausible scenario will it ever resemble its namesake across the Niagara River in Ontario, a tourist mecca with high-rise hotels, theme restaurants, tidy parks, and the 775-foot Skylon Tower. "Niagara Falls is a big shithole," Harry Weist, a retired Goodyear worker, told me as we began a tour of the city one August morning in 2013. "It's really repressed. They got idiots running it." Starting at Interstate 190, we made a counterclockwise loop in Harry's pickup truck: west on Niagara Falls Boulevard, which feeds into Pine Avenue; southwest on Main Street, south on Rainbow Boulevard, and east on Niagara Scenic Parkway, which parallels Buffalo Avenue, once lined with factories operated by companies such as Occidental, Carborundum, and Great Lakes Carbon. The city looked bereft of life, save for the Seneca Niagara Resort & Casino on Fourth Street and a few other hotels and chain restaurants. A half-mile west of the casino, mist rose from the American Falls and, beyond that, the Horseshoe Falls in Canada. "You've got this beautiful attraction," Harry said. "I can't believe you can't do something with this."

We wound up at the Goodyear plant, at Fifty-Sixth Street and Goodyear Drive, where Harry and his father-in-law, Ray Kline, had worked with Rod Halford. Like Halford, Harry and Ray developed bladder cancer. Harry was case No. 37 from the plant, Ray No. 21; both have had recurrences. Harry pointed out the rubber-chemicals division, where all three men had been exposed to ortho-toluidine. Part of the plant's old PVC section had been torn down; the rest was being used as a warehouse. A sign outside read, "TAKE SAFETY TO THE EXTREME. WE MUST. WE WILL." Harry, a man of medium height and build with close-cropped gray hair and a handlebar mustache, was delivering auto parts to supplement his retirement income at the time and regularly passed this spot. "I don't even give it a second thought," he said. We went on to the United Steelworkers Local 4–277 (formerly OCAW Local 8–277) union hall, where we met Harry's best friend, retiree Robert Dutton, and two Goodyear workers. They recounted how bladder cancer—they called it "the ginch" for reasons unremembered—had spread through the plant, ensnaring front-office managers as well as operators and maintenance men. "We had guys that retired, and they said, 'You know what? I worked here all them years, I never got sick, I'm good,'" Harry said. "But that latency period—" Dutton interrupted: "Lo and behold, it bites you in the ass." Harry continued: "That latency period kicks in, and, all of a sudden, they start peeing blood. We all know what happens when you start peeing blood."

Like many postwar factory workers, Harry and Ray earned comfortable livings. Their jobs, while sometimes unpleasant, did not sink to the level of sweatshop labor. They did not toil in dusty coal mines or scorching steel mills. And yet both were ambushed by a preventable, chemically induced illness that could not be blamed on lifestyle, genetics, or rotten luck. Versions of this story have played out around the United States since the mid-twentieth century. The latency period to which Harry Weist referred is the time lag between the first exposure to a toxic substance and the manifestation of disease, a period often measured in decades. Latency is why we still see about 2,500 deaths annually from mesothelioma, a savage cancer of the lining of the lung or abdomen associated with the inhalation of asbestos fibers. Cancer cells invade organs and destroy normal tissue for thirty, forty, even fifty years. Then the malignancy surfaces, dispatching its victim with unusual haste. While workplace conditions overall are not as harsh as they were a half-century ago, membership in unions (the OCAW was instrumental in uncovering the Goodyear cluster and forcing the company to respond) has plummeted, depriving workers of an effective tool to ensure their safety, and the odds of an OSHA inspection on any given day, absent a fatality or serious injury, are infinitesimal. The nation remains ill equipped to protect workers from ancient hazards like asbestos, a fire-resistant mineral spun into blankets and tablecloths by the Greeks, and silica, a ubiquitous mineral that ruined the lungs of stonecutters, masons, and miners millennia ago and is still afflicting construction workers and fabricators of artificial-stone countertops. Modern chemicals such as methylene chloride, found in paint strippers, and trichloroethylene, an industrial solvent, disable and kill because the government refuses to ban them or is slow to restrict their use. Thousands of others may be doing harm; we don't know because we haven't studied them. In 1942, the Industrial Hygiene Foundation of America advised, "Every new chemical or product should be investigated as to its toxicity before it is prepared in large amounts and released to the public." Nothing remotely close to that has happened in the eighty-plus years since.

I began to grasp the scope of this quiet catastrophe while working as a newspaper reporter in Houston, an oil-refining and petrochemical-manufacturing hub, in the 1990s. Far too often there would be fires, explosions, or chemical releases in industrial enclaves such as Texas City, Channelview, or Baytown. Some of these events maimed or killed workers and made front-page news. But the more insidious hazard, I came to learn, was occupational disease. I'd meet workers at union halls and lawyers' offices and hear stories about cancers that ate away skin, lungs, livers, and brains. The plight of these workers was acknowledged obliquely in paid obituaries; one might read that a certain company man died at fifty-five after "a lengthy illness." Decoding was required.

I learned of the Goodyear bladder-cancer cluster in 1998, when I was working on a yearlong investigation of the chemical industry for the *Houston Chronicle*. I made mention of it in a longer article—quoting Rod Halford, one of the twenty-three victims to that point—and put it out of my mind. By 2013, when Goodyear regained my attention amid another investigation, the number of bladder-cancer cases in Niagara Falls had leapt to fifty-eight. I wrote a piece for the Center for

Public Integrity, where I worked at the time. But I could see that the story was too big for a single article—or series of articles. At its core is a fastidious lawyer, Steven Wodka, who represented twenty-eight Goodyear bladder-cancer sufferers over the course of thirty-four years in third-party, product-liability claims against chemical suppliers—the only legal approach that could afford some measure of justice. In none of these cases was there a courtroom reckoning; each was settled without a trial. The corporate defendants wrote checks, insisted that the amounts be kept secret, admitted nothing, and largely avoided publicity. Wodka, fortunately, has an encyclopedic memory and the documents to back it up. And some of his clients—men like Harry Weist and Ray Kline—were willing to share details about conditions in the plant and their experiences with a fiendishly capricious disease. They, Wodka, and others made it possible for me to piece together the debacle that unfolded over decades in Niagara Falls and contextualize it amid the epidemic of disease that is threaded through the American workforce.

CHAPTER 1

NIAGARA FALLS BECOMES AN INDUSTRIAL LEVIATHAN

THE SOUTHERN END OF THE NIAGARA RIVER GORGE, a craggy product of the last Ice Age that lies between the United States and Canada, encompasses three waterfalls collectively known as Niagara Falls. The most spectacular of the three is Canada's Horseshoe Falls, which is 2,200 feet wide at its crest and heaves foaming water into the Lower Niagara River 188 feet below. The American Falls and its adjunct, Bridal Veil Falls, are narrower at the top and shorter in stature. The three give off a constant rumble, dumping 3,160 tons of water over their crests every second. They were born of a phenomenon that began about eighteen thousand years ago, when a layer of ice up to two miles thick covered what is now southern Ontario and western New York. As the ice sheet melted and retreated north, the basins that became the Great Lakes filled with water that flowed toward the St. Lawrence River and the Atlantic Ocean. The falls themselves came into being some twelve thousand years ago, seven miles downriver from their current location; erosion of the bedrock moved them upstream.

Until the mid-seventeenth century, the region was overseen by some twelve thousand members of a tribe dubbed "Les Neutres"—the Neutrals—by French explorer Samuel de Champlain, who had arrived in 1615 and witnessed the tribe's refusal to intervene in skirmishes between the Hurons to the north and the Iroquois to the south. The Neutrals

9

grew crops such as corn, beans, and pumpkins, hunted deer, elk, and beaver, and plucked trout, sturgeon, and salmon from the river. Their bloodless, if not idyllic, existence came to an end in 1652, when many members of the tribe were killed by the Iroquois, who wanted exclusive access to the beaver stocks so they could monopolize the fur trade with the Europeans. Early white explorers gave florid accounts of the area's geography: "Betwixt Lake Ontario and Erie there is a vast and prodigious Cadence of Water which falls down after a surprising and astonishing manner, insomuch that the Universe does not afford its parallel," wrote Father Louis Hennepin, a Belgian priest believed to be the first white man to visit the falls in 1678. "The Waters which fall from this horrible Precipice, do foam and boil after the most hideous manner imaginable, making an outrageous Noise, more terrible than that of Thunder."

The prodigious amount of hydropower bottled up in the gorge was first put to use by a French fur trader named Chabert Joncaire, who opened a sawmill beside the rapids above the American Falls in 1758. Not quite half a century later, in 1805, salt merchants and brothers Augustus and Peter Porter bought a large tract of land in New York that included the falls and founded a village they initially called Grand Niagara, then Manchester, then Niagara Falls. In 1825, the brothers published a flier headlined, "Invitation to Eastern Capitalists and Manufacturers." They trumpeted the region's natural beauty and the sheer power of its waters: "Practically speaking, the extent to which water power may be applied here is without limit. A thousand mills might be erected with the same ease, and equally accessible, as if on a plain; and each supplied with a never-failing water power, at an expence [sic] not exceeding fifty dollars." At first, the response from industry was underwhelming, though a scattering of grist and paper mills did appear by midcentury. "Except in the minds of a few . . . the utilization of any large amount of water from the Niagara River for the generation of power was held to be an undertaking of doubtful wisdom at best," Raymond H. Arnot later wrote in the *Popular Science Monthly*. It wasn't until ground was broken, in 1890, for the Niagara Falls Power Company's first hydroelectric station that the possibilities envisioned by the Porter brothers began to be realized. "By 1893," George R. Shepard wrote years later in the journal *New York History*, "the practically assured success of the Niagara development, assuring the availability of large quantities of cheap power, attracted the attention of chemical and metallurgical industries throughout the country." Some boosters were delusional. In his 1894 utopian tome, *The Human Drift*, King C. Gillette, inventor of the safety razor, foresaw a falls-centered city of sixty million people, which he called Metropolis. "Let us start the ball rolling," he wrote, "with such a boom and enthusiasm that it will draw the wealth and sinew of the nation into its vortex." Nikola Tesla, whose alternating-current electrical supply system was commercially deployed first in Niagara Falls and then in Buffalo, twenty miles away, claimed the gorge's hydropower someday would illuminate the streets of Paris and propel London's trolleys.

Hyperbole notwithstanding, industrial growth around the falls took off toward the turn of the century. Electricity made possible a process called electrolysis, in which electric current triggers a chemical reaction to produce compounds such as chlorine and caustic soda, or to extract aluminum from aluminum oxide. It allowed Edward G. Acheson, a chemist from southwestern Pennsylvania and a former assistant to Thomas A. Edison, to begin making synthetic graphite—a diamond-hard abrasive he called "carborundum," combining in a word the presumed constituents of "carbon" and "corundum," a crystalline form of aluminum oxide—at a factory on Buffalo Avenue in Niagara Falls, New York, in 1895. In fact, a chemical analysis later found, the product consisted of carbon and silicon, but no matter. By the early 1900s, the Carborundum Company was making grinding wheels for automobile and locomotive parts; abrasive paper for the scouring and buffing of shoe soles, wood, and metal; heat-resistant bricks for industrial furnaces; and crystal detectors for radios. The company was wildly successful and profiled glowingly in newspaper articles. "The story of Carborundum is indeed an industrial romance," wrote the *Niagara Falls Gazette*. "It is a romance of fact but it sounds more like the romance of fiction."

By 1906, H. W. Buck, a local electrical engineer "of standing," was boasting in an opinion piece that "the electro-chemical industry, which might be said to have originated through the development of Niagara power, has already reached enormous proportions," but warned that the "exaggerated sentimentalism" some held for the falls could choke off further growth. "The economic side of the Niagara problem is a serious one," Buck admonished in *The Outlook*, "and it cannot be set aside as secondary to that of the scenic interests." Buck needn't have worried. Over the next few decades, factories operated by Hooker, Union Carbide, DuPont, Alcoa, and others sprang up on Buffalo Avenue and adjoining streets. They made compounds such as potash, used in glass, soap, and fertilizer; and acetylene, used in plastics, solvents, and welding torches. Bragged Carbide: "The chemical genie is creating new products faster than the historian can record them." By 1925, promoters were touting the Niagara River as "the greatest single source of water power within a densely populated area in the world [with] . . . a potential of six million horsepower in its rapids and cataracts. . . . To produce such an amount of power by burning coal would require about fifty million tons each year. Imagine 50 carloads of coal coming up to the crest of the falls and dumping the coal over the precipice every minute." Five years later, the city's population had eclipsed seventy-five thousand—some 35 percent higher than it is today.

As World War II loomed, the Buffalo Avenue factories were running flat out, ready to be tapped by government procurers. "The Carborundum Co., with its glowing furnaces of electrically heated crystals, is supplying the nation with greater quantities of abrasive necessities than at any time since 1928," Horace B. Brown wrote in the *New York Times*. "Titanium and Vanadium, the ferro-alloy plants, are operating 24 hour shifts with electric furnaces blazing high into the heavens, mute evidence of the service these purring turbines supply. International Paper turns out the huge rolls of its product. The Shredded Wheat factory ships away more cartons of its cereal." On December 5, 1941, two days before the Japanese attacked Pearl Harbor, the *Gazette* heralded the opening of a DuPont plant built by the government to supply "defensive chemicals," such as poison gas–absorbing activated charcoal, for the army's Chemical Warfare Service. By 1942, almost half of DuPont's gross sales came from products that hadn't existed or weren't manufactured on a commercial scale a mere fourteen years earlier. Scores of these chlorine- and sodium-based compounds were helping "to impart armor toughness to resist the impact of enemy shells, to prime gasoline with the added power that is giving Allied planes superiority, both in Europe and the Pacific, and to provide life-saving drugs that help the wounded," DuPont executive F. L. Livingston said in a speech to plant foremen in February 1945.

At the time Livingston spoke, DuPont and its neighbors were begging for workers and unable to meet wartime production schedules despite the companies' best efforts. "For two years," according to the *Buffalo Courier-Express*, "Niagara Falls employers have pooled interviews and staged labor recruitment drives in various sections of New York State and in Alabama, Kansas, Ohio, Pennsylvania, Minnesota, Mississippi, Missouri, Texas, West Virginia and Wisconsin." These outreach efforts had proved largely fruitless. "Plant executives pointed out that anyone looking for a job should not be scared away by technical-sounding chemistry terms," the newspaper reported. "They said any person of average intelligence can be trained easily to do one of the specialized jobs in the industry." Photos accompanying the article show two men operating "the largest abrasive wheel press in the world" at Carborundum and five "girls" in the Hooker lab working with chlorine, caustic soda, and hydrogen products.

Within six months, the war was over. Goodyear, loath to miss out on the industrial bacchanal in Niagara Falls, announced plans to build a vinyl factory near the corner of Fifty-Sixth Street and what was then Baker Avenue. A short *Gazette* article explained that the plant had been "formerly scheduled for construction at Natrium, W. Va., but was moved to New York because of convenient availability of raw materials and lower cost of plant investment." At the time, Goodyear, headquartered in Akron, Ohio, was nearing its fiftieth anniversary and was known globally for its tires and its silver, blue, and yellow blimps. The company had been founded in 1898 by Frank A. Seiberling, a squat, irrepressible native of the Akron area who was then thirty-eight and nearly broke. His father, John, had run an assortment of milling, banking, and manufacturing enterprises before losing most of them in the depression of the 1890s. During a chance meeting with a business acquaintance in Chicago, Frank agreed to buy an abandoned strawboard factory—a sad complex that straddled the Little Cuyahoga River in East Akron—for $13,500. Seiberling had "some doubts about his ability to pay for it, and no definite ideas on what to do with it," according to a manuscript prepared years later by Goodyear's in-house historian. He borrowed $3,500 from his brother-in-law and—having loosely followed the operations of his father's Akron India Rubber Company, sold the year before—decided to go into the tire and rubber business with his brother, Charles. The move was not without logic: Americans were enamored with bicycles, and the horseless carriage—the automobile—had recently come on the scene. But it was far from a sure thing.

Seiberling named the new company after Charles Goodyear, a Connecticut-born inventor who had discovered the process of vulcanization—the treatment of natural rubber with sulfur, heat, and pressure to harden and strengthen it—and died penniless. Goodyear chronicled his triumphs and travails in a thick 1853 volume with an unappealing title: *Gum Elastic and Its Varieties, with a Detailed Account of its Applications and Uses, and of the Discovery of Vulcanization.* A fifteen-thousand-word critique published in the July 1865 edition of the *North American Review* noted the book's vapidity ("written without art or skill") but admired its lack of guile and unique physical composition: its cover and 620 pages were made of vulcanized India rubber. The former resembled "highly polished ebony," the latter "ancient paper worn soft, thin, and dingy by numberless perusals." The *Review* article paints a sympathetic picture of Goodyear's life—how a "bankrupt hardware merchant" then living in Philadelphia became obsessed with a commodity made from the milky sap of trees in the tropics of Asia and South America after visiting a depot in New York operated by the Roxbury India Rubber Company in 1834. The nascent company, out of Massachusetts, had thrived by selling rubber shoes and fabrics during the cooler months of the year, when they would not melt. The feint, however, did not last; the products disintegrated in warmer weather, and customers demanded their money back. The rubber market collapsed, and by the end of 1836 the company was no more.

Goodyear was certain he could find a solution to the problem, though he had many handicaps: scant knowledge of chemistry, an aversion to mathematics, poor health, and grinding poverty. At one point, according to the *Review*, "he was arrested for debt, and compelled to reside within the prison limits. He melted his first pound of India rubber while he was living within those limits and struggled to keep out of the jail itself. Thus he began his experiments in circumstances as little favorable as can be imagined." Goodyear nonetheless had a "conviction that India rubber could be subjugated, and that he was the man destined to subjugate it." He began a series of experiments at a small house in Philadelphia in the winter of 1834–35, melting the sap, kneading it, and rolling it into sheets that appeared heat-resistant at first but turned to "soft and stinking paste" by summer. He solicited advice from chemists and physicians but learned nothing useful. Later, in New York, he tried mixing the gum with magnesia and boiling it in a solution of quicklime—a caustic white alkali—and water, producing sheets of rubber so smooth and firm that they won accolades in the press and sold briskly. Then he discovered that a mere splash of vinegar or another weak acid made the cloth sticky and ruined the effect. The failed experiments went on until Goodyear treated the rubber with aqua fortis, a concentrated form of nitric acid that contained what turned out to be the magic ingredient: sulfuric acid, which mitigated the effects of heat. It was an early breakthrough.

A brief period of hope was dashed by a banking crisis that became known as the Panic of 1837, which, according to the *Review*, "reduced poor Goodyear to his normal condition of beggary." His family and friends pleaded with him to return to the hardware business. But he persisted with his experiments until, in a fit of pique one spring day in 1839, he knocked a sulfur-infused blob onto a hot stove and saw that it charred but did not melt. Two more years passed as Goodyear sought to perfect his curing technique, borrowing his wife's stove and saucepans to bake and boil the goo. Broke as usual, he landed again in debtor's prison. All told, it took him more than ten years to produce, as the *Review* put it, "a new class of materials, applicable to a thousand diverse uses. It was a cloth impervious to water. It was paper that would not tear. It was parchment that would not crease." He all but gave away the rights to manufacture India rubber shoes; by 1865 a million and a half pairs were sold annually. Patent infringers nibbled away at him, selling rubber engine belts, artificial teeth, and stretchers for the battlefield. "No inventor probably has ever been so harassed, so trampled upon, so plundered by that sordid and licentious class of infringers known in the parlance of the world, with no exaggeration of phrase, as 'pirates,'" wrote Joseph Holt, US commissioner of patents from 1857 to 1859. "The spoil of their incessant guerilla warfare upon his defenceless [sic] rights have unquestionably amounted to millions."

Goodyear died a pauper in a New York hotel at the age of fifty-nine on July 1, 1860, and was buried in his hometown of New Haven, Connecticut. The company named after him was incorporated on August 29, 1898; Frank Seiberling chose as its logo the winged foot of Mercury, the ancient Romans' messenger of the gods. In its first year the Goodyear Tire & Rubber Company generated profits of $34,621 from sales of bicycle and carriage tires, horseshoe pads, rubber bands, and poker chips. The bicycle tires it made were cheap—"unguaranteed" in the parlance of the time—at a cost to the company of forty-eight cents apiece. "And, of course, a tire that could be made for 48 cents didn't contain too much rubber and was mostly compound, and it wasn't worth a whoop," Seiberling said in a speech some years later. Assigned the unfortunate brand name Tip Top, the tire routinely popped off the rim and sent the rider tumbling. "All over the country we heard about those 'rotten Goodyear tires.' . . . The stain of that Tip Top Unguaranteed Bicycle Tire followed us for at least ten years," Seiberling said. "If there is one thing that the Goodyear Tire & Rubber Co. has learned as a result of this bitter experience, it is that you can't make an inferior article—a poor-quality article—and live."

Seiberling made his most important hire in 1900, when he brought on Paul Litchfield, an MIT graduate with technical training in tire-making, as superintendent of the Akron factory. The two could not have been less alike: Seiberling, according to the official corporate history, was a "short, dynamic, daring, and mercurial opportunist of great persuasive power." Litchfield was "tall, dignified . . . courtly, and analytical." It was Seiberling who invented the tire-building machine, the straight-sided tire, and the detachable rim, and brought the lighter-than-air dirigible to America (the 258-foot, hydrogen-filled *Akron* was assembled in 1910 and exploded twenty-three minutes into its first transatlantic flight in 1912, killing the crew of five). By 1916, Seiberling had built Goodyear into the world's largest tire company; its sales exceeded $100 million the following year. Falling prices and bloated inventory, however, forced him to hand over control of the company to a group of bankers during the "little Depression" of 1921. Not one for self-pity, he founded Seiberling Rubber Company in the Akron suburb of Barberton, Ohio, later that year and served as its board chairman until he was ninety.

With Seiberling out of the picture, Litchfield became the face of Goodyear and was named president in 1926. In an article marking the company's thirtieth anniversary two years later, Litchfield observed that rubber manufacturing had become a billion-dollar enterprise worldwide, with Goodyear as its undisputed leader. This he attributed not only to the quality of the company's tires, rubber heels, engine belts, and aircraft skins but also to the character of its thirty-two thousand workers. "Goodyear's personnel policy . . . has been to select its men carefully, to train them thoroughly, to pay them a little better than the going rate of wages, to create good living and working conditions so as to attract and hold the better type of men, and to give definite encouragement to initiative and opportunity," Litchfield wrote. "An intelligent, adequately paid and stable force of men will produce a better product and do it far more economically than a low-wage, fluctuating force with resulting high labor-training costs, waste of time and material and faulty workmanship." By this point he had introduced the Goodyear Industrial Assembly, a company-sanctioned union, to give rank-and-file employees some measure of influence over corporate policy, and Goodyear Industrial University to encourage their intellectual growth. The future, Litchfield predicted, was exceedingly bright: "The saturation point for automobiles has not yet been reached in this country; general use of the automobile abroad has scarcely begun; tire requirements for the rapidly developing truck and bus transportation will be enormous; aerial transportation is yet in its infancy . . ." A photo gallery featured Goodyear tire and fabric plants in Sydney, Toronto, and Los Angeles, among other places.

Litchfield carried on Seiberling's quixotic passion for giant, rigid airships. The company built the *USS Macon* and a second iteration of the *Akron* for the US Navy at the cavernous Goodyear Airdock, only to see a series of disasters smother the industry. The helium-filled *Akron* went down in a thunderstorm off the New Jersey coast in 1933, killing all but three of the seventy-six men aboard; among the lost was Rear Admiral William A. Moffett, chief of the navy's Bureau of Aeronautics. The *Macon* flew for two more years before plunging into the Pacific in foul weather off California; seventy-nine of the eighty-one aboard survived thanks to a timely SOS call. When the German Zeppelin *Hindenburg* crashed spectacularly in flame before newsreel cameras and took thirty-six lives while landing in Lakehurst, New Jersey, on May 6, 1937, the era of the passenger airship was finished. Deeply disappointed, Litchfield redirected his energy toward the improvement of smaller, helium-filled promotional blimps, modern iterations of which still fly above football stadiums, golf courses, and other venues. He oversaw the opening of tire plants in Argentina and Java and the acquisition of a rubber plantation in Costa Rica. With the Industrial Assembly's blessing he instituted a six-hour workday in Akron to keep as many employees on the payroll as possible during the Great Depression.

Forward-thinking though he might have been when it came to employee relations, Litchfield could not sidestep the turbulent labor climate of the 1930s. With ardently pro-union Franklin D. Roosevelt in the White House, the American Federation of Labor began a drive to organize workers at Goodyear and two other rubber-industry titans, Firestone and B.F. Goodrich, in 1933. Two years later, the National Labor Relations Act gave a legal boost to the organizers, weakening the grip of company unions such as the Industrial Assembly. A strike, ostensibly over working hours but primarily aimed at establishing the United Rubber Workers of America as the official bargaining unit at Goodyear, commenced in February 1936. Over five weeks in Akron, a few hundred members of URW Local 2 were joined on the picket lines in snow and subzero cold by thousands of their brethren from other rubber companies and a contingent of Appalachian coal miners dispatched by John L. Lewis, the charismatic leader of the United Mine

Workers of America. Litchfield and other Goodyear managers encamped in the plants to keep them running in some limited fashion as fourteen thousand workers were idled. A union newspaper, the *United Rubber Worker*, heralded the picketers for their "courage and tenacity"; "Heroes Fight Wintry Blast During Strike," one headline read. Although the URW won concessions (but not formal recognition) in a settlement with the company, the unrest continued for months. An article in the *American Mercury* by managing editor Gordon Carroll offered a mordant recap: "On ninety-four separate days since March 21, 1936, the official settlement of the strike, sit-downs or flagrant acts of intimidation or assault have been recorded. Non-union members have been slugged and beaten; their homes have been picketed and, in some instances, bombed; riots have been staged at company gates; a general state of confusion and enmity has been maintained." The URW called the strike "an unforgettable event in labor" and blamed the post-settlement turmoil on Goodyear supervisors and Industrial Assembly stalwarts who "poked fun" at the terms of the agreement.

The company endured another walkout—brief but violent, involving the use of tear gas and riot guns by Akron police, with more than a hundred injured—in 1938, and a twenty-day stoppage in June and July of 1945 that prompted the seizure of five Goodyear plants by the navy on President Harry Truman's orders. In a "message to employees" published in the *Akron Beacon Journal* on August 22, 1945, a week after V-J Day, Litchfield was effusive. Even before Pearl Harbor, he wrote, Goodyear factories had churned out "specialized war products for the anti-Nazi nations." (The company's prewar reach was vast indeed: by the time Adolf Hitler invaded Poland on September 1, 1939, Goodyear was selling its wares on every continent but Antarctica. It had eighteen subsidiaries, seven factories, seven rubber plantations, thirty-seven branches, twenty-eight depots, and hundreds of distributors outside the United States.) For the duration of the war, "Goodyear research and development led the way and men and women of Goodyear production lines followed up with a stream of new products—fuels cells, half tracs, boats, barrage balloons, specialized tires, gun mounts, camouflage material, gas masks and a host of others," Litchfield wrote. "That stream became a torrent; never have we produced such huge tonnages." Now it was time to turn to peacetime production: "The public wants our tires, tubes, mechanical goods, rims, Airfoam [cushioning], Pliofilm [packaging material] and other products as quickly as they can be made." Litchfield acknowledged URW Local 2 as the Akron workers' bargaining unit and promised Goodyear management would "go more than half way to make the relationship between the company and the union a pleasant and effective one."

In 1946, Goodyear recorded sales of $617 million—nearly twice what it had brought in at the start of the war—and moved aggressively into the manufacture of chemicals and plastics. In November of that year a Goodyear subsidiary, Pathfinder Chemical Company, opened a plant in Niagara Falls that made vinyl chloride, a sweet-smelling gas, and turned it into polyvinyl chloride (PVC) resin, a fine, white powder. Goodyear called its version Pliovic. The powder was loaded into fifty-pound paper bags and shipped to fabrication plants around the country, where it was heated and molded into wire insulation, flooring, shoe soles, raincoats, shower curtains, and toys. The first such resin had been produced more than a century earlier, in 1838, by German chemist Justus Freiherr von Liebig and his French student, Henri Victor Regnault. But it was not commercialized in America until the 1930s, when B.F. Goodrich introduced what it called "non-rigid vinyl chloride plastics." The US Department of Commerce was bullish on PVC after the war, describing the resin as "tough, odorless, tasteless [and] noninflammable" and the products it yielded as having "high resistance to warping, moisture absorption, cold flow and shrinkage." Vinyl, the department said, was "one of our most versatile plastic families." Goodyear wanted in. In a speech to the Cleveland Chamber of Commerce in October 1945, R. P. Dinsmore, Goodyear's vice president for research, regaled his audience with a list of vinyl's uses: as a liner for beer cans and bottle caps, a flameproof coating for fabric and wire, and a finish for paper labels, metal tubes of ointment, and rubber balls "to enhance appearance and sales appeal as well as to afford resistance to oil, grease, alkali and water and facilitate cleaning."

There were three problems with vinyl chloride—a monomer, or molecule, that can bond in long chains to make a polymer such as PVC. One, it made workers giddy as drunks if they inhaled enough of it while cleaning the reactors in which the gas was cooked under pressure and turned into resin. Two, it was combustible. And three, studies later would find, it could cause cancer.

CHAPTER 2

RAY AND DOTTIE

B Y JUNE 1940, as hostilities simmered in the Pacific, Goodyear and other manufacturers that relied on natural rubber faced a grim possibility: imports of the white sap, known as latex, from Southeast Asia were likely to be curtailed, perhaps severely. In response, President Roosevelt ordered the formation of the Rubber Reserve Company—the RRC—which coordinated the stockpiling of the precious gum and the collection and recycling of tires and other scrap. It would not be enough; by the end of 1941, the four big rubber companies had lost access to 90 percent of their supply. The RRC handed them a formidable assignment: ramp up production of synthetic rubber to 400,000 tons annually, from a few hundred. (At the time, the United States used about 600,000 tons of natural rubber a year. A tank required a ton, a battleship 75 tons.) Synthetic rubber was largely unproven: the Germans had cooked up small amounts of it during World War I, but it was of poor quality. In a show of cooperation that seems astonishing today, the four companies agreed on a recipe, based on the monomers butadiene and styrene, by the spring of 1942. Production that year fell short of 4,000 tons; by 1945, it had reached nearly 720,000. Natural-rubber supplies again were disrupted in Indonesia and what was then Malaya at the start of the Korean War in 1950, and Truman signed legislation that put the synthetic-rubber factories under government control until 1955. The country never reverted to its dependence on natural rubber, and manufacturers sought to improve the man-made version so that, for example, tire treads wouldn't crack from the destructive process of oxidation. In September of 1958, a Goodyear press release touted a new antioxidant: Wingstay 100, a "blue-brown flaked solid" known internally as Nailax. It was made at the company's plant in Niagara Falls, in Department 245—rubber chemicals. Its flakes compared in size to rolled oats. Among its ingredients was a pale-yellow chemical called ortho-toluidine, part of a family known as aromatic amines, which had begun spawning bladder cancer in synthetic dye workers in Europe in the late nineteenth century.

At the time of Goodyear's announcement, Raymond Walter Kline was twenty years old, newly married and in search of steady employment. He'd graduated two years earlier from Beccaria-Coalport-Irvona (BCI) High School in Clearfield County, Pennsylvania, a hundred miles northeast of Pittsburgh. The only child of Chester and Dora Kline, he was born on July 15, 1938, in Irvona, a borough of one thousand people underlain by bituminous coal deposits. His father worked in a brickyard, making furnace liners for steel mills, and had to quit at fifty-five because of heart trouble. His mother worked at a five-and-dime store and later cleaned the offices of a coal company. The girl Ray would marry—Dorothy Mavis Troxell, who goes by Dottie—lived in the area on and off during her youth. If Ray's early life was mostly uneventful—he hunted deer, fished, and played trumpet in a dance band and at military funerals—Dorothy's was cartoonishly awful. Her fragile, trusting mother, Bertha, was twice married, both times to abusive men. The first pushed Bertha out of a moving car. When she developed cervical cancer, the second wouldn't let her go to the doctor until the cancer had metastasized. She died on August 24, 1953, two months after moving to Newfane, in western New York, where her brother and sister lived. "She was fair game," Dottie said, "for all the louses out there." Dottie, living with her maternal grandparents, met Ray Kline in 1954, during a play rehearsal her sophomore year at BCI High. The two began dating; Dottie thought Ray resembled James Dean, a comparison he did not discourage. She knew him as Sonny, a nickname he acquired from his parents. She still addresses him as such, and remains the more unfiltered member of the couple, speaking her mind on all subjects while Ray mostly remains silent. The abuse her mother passively endured steeled Dottie's resolve not to be a pushover.

Ray's first job out of school in 1956 was as a truck driver, hauling coal from open-pit mines to a rectangular structure called a "tipple," from which railcars were loaded. Tiring of running coal for meager pay, Ray cycled through four jobs in 1957 alone: testing fish cans at the Conneaut Can Company in Conneaut, Ohio; assembling car doors at Ford Motor Company's Woodlawn plant in Buffalo (he was fired for skipping work during deer season); operating a punch press at the Auto-Lite battery plant in Niagara Falls; and transferring wheat from railcars to the silos at the National Biscuit Company, or Nabisco, also in Niagara Falls. Ray—restless, not yet twenty—yearned for steady work and was beginning to think of settling down. He and Dottie married on February 22, 1958, and returned to Pennsylvania after he was laid off by Nabisco. He was introduced to the chemical industry the following November, when he applied for work at the Olin Mathieson plant in Model City, New York, a failed utopian community outside Niagara Falls whose promoter, William T. Love, oversaw the aborted construction of an eponymous canal later acquired and filled with toxic waste by Hooker Chemical. Ray was hired as a chemical operator in the plant's K area, where fuel for the short-lived B-70 nuclear strike bomber was made. He took samples of the liquid mixture during the production process and was required, under penalty of discharge, to wear a canister-style, full-face respirator to protect against a pungent, sleep-inducing gas the workers knew only as "Goodies." After all the moving around, Ray and Dottie had made Niagara Falls their home, and their first child, Diane, was born at Niagara Falls Memorial Medical Center on December 7, 1958. The Olin plant closed a year later, but Ray's unemployment was brief. He landed a job at Goodyear in January 1960 for a respectable $1.36 an hour.

The vinyl side of the Goodyear plant, known as Department 145, was about to undergo a million-dollar expansion that would double its production capacity, from twenty million to forty million pounds of resin per year. Ray filled bags with Pliovic and cleaned the guts of the PVC reactors in Building E-1, chipping the hard, white residue off the sides with a hammer and chisel and frequently getting dizzy from the vinyl chloride fumes. Dottie, who had been staying home, taking care of Diane, bore a son, Mark, on September 11, 1961, and the young family moved into a small house on Seventy-Second Street, living comfortably on Ray's wages (a gallon of regular gas cost 31 cents, a gallon of milk a half-dollar). Niagara Falls and Buffalo, meanwhile, were thriving. A promotional pamphlet published in 1963 plugged the region as "the largest electro-chemical and electro-metallurgical center in the world. . . . Of the largest 500 U.S. manufacturing companies, 22 have plants here. The area labor force numbers over 500,000 skilled in manufacturing, research and service industries and are easily trained for any industrial, commercial or research operation."

One day in June of that year, during a visit to Pennsylvania, Mark, not quite two, became fussy and feverish. His parents thought he was simply cutting teeth, but the fever spiked, and they took Mark to the hospital. He died within hours of spinal meningitis. Dottie was hysterical and had to be sedated. Two aunts packed away Mark's crib, toys, and clothes to spare Dottie the trauma. Diane, then four, had shared a room with her little brother and was racked with nightmares after his death.

The loss of one child is enough to break the strongest parent. Ray and Dottie would be tested twice more in short order. On April 11, 1964, John Andrew Kline was born without much of his brain and skull, a condition called anencephaly. He lived less than a day; Ray wrapped the baby in a blanket, put him in the back seat of the car, and drove him to Pennsylvania to be buried. On May 18, 1965, Dona Marie Kline was born with brain-fluid buildup, known as hydrocephalus, and spina bifida, a spinal cord defect, and died after thirty-three days. Decades later, Dottie would say, "Had I not known the Lord, I would have been institutionalized." The sting of these losses stayed with her, and her suspicions eventually fell on Ray's work around vinyl chloride, which some studies had linked to birth defects. "Do you know what it's like when they have to place a [hospital] mat over your face when your baby's born so you won't see it?" she asked. "Do you know what it's like to carry a baby for nine months and have it die at six weeks old? Those are the things I remember about Goodyear."

There's no way to say with certainty that Ray's workplace exposures contributed to the Kline children's injuries. (Dottie gave birth to a healthy son, Raymond, in 1966.) But it's clear that Ray took in vast amounts of vinyl chloride at Goodyear. Early in his nine-year stint in Department 145, he experienced numbness in his fingers, caused by a degenerative bone condition called acroosteolysis. The plant doctor told him it was a side effect of cleaning crud from the reactors. He never used gloves, nor did anyone recommend that he do so. When Ray transferred back to Pliovic packaging, the condition went away. But acroosteolysis had become an industry-wide problem.

It had been discovered in 1964 by Dr. John Creech Jr., a physician at B.F. Goodrich's PVC plant in Louisville, Kentucky. Like Ray Kline, the ailing Goodrich workers had been exposed to hundreds, if not thousands, of parts per million of vinyl chloride while cleaning PVC reactors. Creech gave a presentation on the phenomenon at an American College of Surgeons meeting that year in Louisville, having responded to an appeal from organizers seeking "bizarre cases" for discussion—"anything but hernia repairs and varicose veins and hemorrhoids." Goodrich made no attempt to stop him; in fact, it reported the hand problem to the US Public Health Service. But things were going on behind the scenes—things Creech didn't know about until late in his life.

By early 1965, the Goodrich acroosteolysis cluster was being discussed with apprehension by other PVC manufacturers and their trade group, the Manufacturing Chemists' Association, or MCA, precursor to the Chemical Manufacturers Association. Two officials with the chemical giant Monsanto took a fact-finding trip to Goodrich corporate headquarters in Akron in early 1966 and learned, according to their trip report, that twenty-two Goodrich employees were suffering from the problem. All but one cleaned reactors. Goodrich agreed to share its data with the other companies and read their employees' X-rays free of charge but hoped, according to minutes of the MCA's Medical Advisory Committee in late 1966, that "all will use discretion in making the problem public."

In January 1967, the MCA commissioned an epidemiological study of PVC workers by the University of Michigan's Institute of Industrial

Health—and then altered the study design to make sure it wouldn't un-
earth anything too damning. Workers were classified only by job title,
not by chemical exposure; predictably, no individual chemical—vinyl
chloride, for example—was implicated. The planned research cohort
of twenty thousand wound up being closer to five thousand—thereby
diluting the study's power—owing to flagging participation by the companies. Their representatives, among them the Goodrich and Goodyear
medical directors, found objectionable even the watered-down report
delivered to them in confidence by the Michigan investigators in 1969.
Why? Because it hinted that acroosteolysis might be indicative of systemic toxicity, a much bigger worry than stubby, deadened fingers, and
proposed that vinyl chloride exposures be limited to 50 parts per million,
a recommendation ordered deleted from the final report. The Michigan
team wanted to do a follow-up study. The MCA passed, ensuring that
uninformed workers would continue to be exposed to vinyl chloride
fumes. It was around this time that Ray Kline left the vinyl side of
Goodyear to become a plant-wide mechanic. The move increased, rather
than lowered, his cancer risk, though he didn't know it. It put him in
regular contact with a new carcinogen: ortho-toluidine.

CHAPTER 3

AN AMERICAN
"CASUALTY LIST"

O BTUSENESS TOWARD WORKPLACE health and safety risks—or willful denial in some cases—was not uncommon within American industry in the 1960s. Production was key; everything else was subordinate. In a 1965 report to US surgeon general William Stewart, a panel led by William Frye, chancellor of the Louisiana State University Medical Center, called for a national program whose goal would be "elimination or control of any factor in the work environment which is deleterious to the health of workers." The US Public Health Service should be given "specific legislative responsibility and necessary resources" to launch such a program, which would "reduce work absence, increase productivity, and strengthen the economy," the so-called Frye Report said. It added: "The most pressing need is for criteria for safe exposure to chemical contaminants." At the time, the principal federal law governing health and safety on the job was the Walsh-Healey Public Contracts Act of 1936, which applied to businesses that did a certain amount of government work. Few pretended it was effective; there were only fifteen inspectors to enforce it nationwide. The Frye Report set in motion a rush of activity within the Johnson administration. The Public Health Service circulated a brochure declaring that, given estimates that there were at least 336,000 cases of preventable illness among America's seventy-five million workers each year, the moment demanded a "massive effort" by government, industry, and science to vanquish the epidemic. The brochure included a photo of a worker wearing no protective gear and standing expressionless amid mounds of asbestos in a dust-clouded room; he was among 3.5 million at risk of dying from the lung-choking fibers, the agency said. The brochure summed up other dangers in bullet points: bladder cancer among dye workers, black lung among coal miners, lung cancer among uranium miners, skin cancer among roofers who used coal tar pitch. And, it warned, "the possibility of genetic damage, affecting generations yet unborn, is an immeasurable, but very real, threat whenever the living body is subjected to a constant barrage of different chemical and physical assaults."

President Johnson himself emphasized the magnitude of the crisis in a speech to a group of union newspaper editors in May of 1966. "We do not know the full, long-range impact of these new hazards on the health of our workers," he said. "We do not know enough about what really happens to men and women who handle chemicals, plastics, asbestos, petroleum products and glass. We do not know enough about the effects on a worker subjected to extremes of heat, cold, noise or humidity. Despite all the research we have done, we do not even know the full effects of radioactivity." In its account of the speech, the *New York Times* reported that Johnson promised "with a wry grin" to press on with legislation even though, the president said, "the polls sometimes indicate that we're moving ahead too fast." In a confidential memo to White House domestic adviser Joseph Califano in December of that year, the deputy director of Johnson's Office of Science and Technology, Ivan Bennett Jr., said that attempts to shore up the anemic, state-run system of workplace-safety regulation then in place had been "totally unsuccessful," imperiling millions of workers not covered by Walsh-Healey or other toothless federal statutes. The situation had been "compounded by a lack of Federal leadership (Congress and Executive Branch), public apathy, industry opposition, little real support from organized labor (in contrast to their efforts with respect to medicare) and little or no effective action at the State level," Bennett wrote. A federal response began to take shape in 1967. Secretary of Labor Willard Wirtz outlined it in an "eyes-only" memo for Califano at year's end: legislation that was being readied for introduction in January would give the Department of Labor the authority to set and enforce safety and health standards, and give the Department of Health, Education and Welfare primary responsibility for research. States would be awarded grants "to encourage them to protect that part of the work force within their reach and outside the Federal reach."

On February 1, 1968, two days after the Viet Cong had stunned war-fatigued America by launching what became known as the Tet Offensive, the House Select Subcommittee on Labor opened the first of eleven hearings on a crisis as repugnant and terrifying in its own way as the bloody assault on thirteen South Vietnamese cities. The subject: "A bill to authorize the Secretary of Labor to set standards to assure safe and healthful working conditions for working men and women." Wirtz, the inaugural witness, asserted that just as "this morning's paper reports a casualty list from Vietnam," the lawmakers should be made aware of "a casualty list repeated every single workday throughout the year in this country"—sixty killed and nine thousand hurt in workplace accidents. On an annual basis, 14,500 were dying and more than 2.2 million disabled—numbers that did not include the hundreds of thousands felled or incapacitated by occupational illnesses, which could take decades to make their presence known. At least $1.5 billion in wages—$10.8 billion in today's dollars—were being lost by the injured workers. And yet, Wirtz testified, it was "almost impossible to realize the under-emphasis which has been placed on this matter over the years." Government spending on occupational safety and health research amounted to $6.57 for each work-related death, compared with $95 per traffic death. Some states were doing a serviceable job of regulating the workplace, but many weren't even trying. Heavily industrialized Ohio, for example, had 109 wildlife inspectors but only 79 workplace-safety inspectors. Mississippi had no labor department and two safety inspectors. When Pennsylvania banned the use of beta-naphthylamine, one of the most carcinogenic of the aromatic amines, a company that made the chemical there relocated to more lenient Georgia. The upshot of this neglect and regulatory inconsistency was an unrelenting stream of macabre wounds—severed arms and fingers, crushed skulls, burns that ate into muscle—and diseases that consumed the lungs, liver, and brain, among other organs. "The problems are getting worse, not better," Wirtz said. Since passing a bill in 1913 to address "frightful diseases" among workers who made phosphorus matches, Congress had done almost nothing to reduce "the human costs—in terms of preventable death, disease, injury—of doing business."

At the ten subsequent hearings, members of the subcommittee heard from dozens of other witnesses, among them Surgeon General Stewart, who summarized a study of 1,700 factories employing 142,000 workers in six metropolitan areas in 1966 and 1967. Sixty-five percent of the workers were "potentially exposed to toxic materials or harmful physical agents, such as severe noise or vibration." Yet only 25 percent were "protected adequately" from these hazards, in the researchers' estimation. A more recent study of 231 foundries in one state, which Stewart did not identify, concluded that one in fifteen workers was exposed to "environmental conditions which were capable of producing disabling and fatal diseases." Foundries were notorious for exposing their workers to dangerous levels of silica dust, which embeds in and scars the lungs and slowly suffocates its victims. But the disease known as silicosis no longer was confined to the "dusty trades," Stewart cautioned. Silica exposures were "so widespread in industry today" that an estimated four million workers were at risk of illness. Meanwhile, workers were "coming in contact with literally hundreds of new chemicals and formulations in industrial uses every day. . . . Today the chemicals have increased so fast in number that we have recommendations for only 400 out of 6,000 in commercial use."

Tony Mazzocchi, the spirited legislative director of the Oil, Chemical and Atomic Workers International Union, presented the subcommittee with a catalogue of horrors suffered by union members: seven killed in an explosion at the Atlantic Richfield refinery in Port Arthur, Texas; seventeen workers at the Phillips refinery in Kansas City, Kansas, dispatched by leukemia or lung cancer in the previous twelve years; "abnormally high incidence of heart disease" in the past five years at Gulf Oil in Cincinnati. Mazzocchi introduced Adolph Wisnack, president of the OCAW local representing employees of the National Lead

Company in Sayreville, New Jersey, where a mysterious scourge had claimed three lives. At first, workers at National Lead had complained of severe headaches and taken to carrying bottles of aspirin and Bufferin in their pockets. "I became more alarmed," Wisnack testified, "when I heard one of the employees in his early twenties [had died] of a cerebral hemorrhage." A few months later, the same condition claimed a "very athletic employee," twenty-five. Just weeks before the hearing, a third man, also twenty-five, had succumbed. The culprit turned out to be colorless, odorless carbon monoxide, levels of which were far above recommended limits. Inspectors with the state of New Jersey, which supposedly had one of the better safety programs, had visited the plant but failed to detect the problem, Mazzocchi said. National standards and enforcement were desperately needed, and not just to protect workers. "We have workers killed on the job, but we are dealing with situations where, if there is an accident, we have it within our means to kill tens of thousands of people," Mazzocchi said. "Something must be done. I want to reiterate that a worker, who, dulled by carbon monoxide in small doses, . . . and is involved in an accident that releases chlorine or anything else, is endangering the entire community. Modern society cannot take the attitude that what happens in the particular facility is the business of the people in that facility. It is the business and the concern of the community at large."

Irving Selikoff, a physician and professor of community medicine at the Mount Sinai School of Medicine in New York, testified that he had spent the previous six years studying the incidence of disease among insulation workers in the construction industry. One of the materials with which they worked—asbestos—had been "under suspicion as a health hazard for some time," Selikoff said in a measured but forceful tone, but implementation of safety recommendations had been "haphazard and inadequate." As a result, mesothelioma—a tumor once so rare it didn't have its own classification on an international list of causes of death—had become common among insulators whose lungs filled with asbestos fibers. Thirteen of the last 113 deaths at one local had been attributed to mesothelioma of the pleura—a thin layer of tissue that covers the lungs—Selikoff said. "In other words, one out of 10 instead

of one out of 10,000," which had long been the incidence rate. "Tell
me what has happened in the industry to improve conditions over the
past years, over the years you have been intimately associated with the
problem," asked subcommittee chairman James O'Hara, a Democrat
from Michigan. "Mr. O'Hara," Selikoff replied, "if very much had
happened, I wouldn't be here."

Industry leaders—who, for the purposes of these hearings, at least,
were fierce states-rights advocates—had their turn before the subcommittee. They saw the legislation as a naked power grab by the federal
government and were aghast at the prospect of being babysat by a new
agency that could levy civil penalties of up to $1,000 and criminal pen-
alties of up to $5,000, with possible prison time thrown in, and issue
cease-and-desist orders that could shut down businesses. What, pleaded
Leo Teplow, vice president of industrial relations for the American Iron
and Steel Institute, was the "great rush" to usurp the states? (At a Senate
hearing in June, Teplow would take umbrage at an "inflammatory, lurid"
booklet released by the Labor Department to muster public support for
the legislation. The cover of *On the Job Slaughter: A National Shame*
showed a bloodied forklift driver crushed to death by bales of what appears to be cotton. Inside were images of a severed hand next to a power
saw and a lifeless ironworker buried in heavy beams.) Raymond Lyons,
a vice president with the Fruehauf Corporation speaking on behalf of
the National Association of Manufacturers, assured the subcommittee that industry was more than capable of self-policing, noting that
"people are estimated to be 10 times safer at work than they are away
from work." Lyons said his organization "strongly questions whether
the imposition of Federal standards will materially contribute to better
occupational safety and health conditions." DuPont's safety director, J.
Sharp Queener, testifying on behalf of the US Chamber of Commerce,
complained about "punitive, not constructive" enforcement measures
proposed for errant employers—a minority in the chemical industry, in
his view—and insisted that workers themselves bore responsibility for
their own well-being.

Queener praised DuPont for its decision to stop making beta-
naphthylamine in 1955. "It is good business to know that you are not

putting into the public market a product that is really too hazardous to justify it," he said. Pennsylvania banned the chemical, O'Hara noted, because it was "known to produce bladder cancer in workers known to be exposed to it." Was this why DuPont did away with it? "Yes, we knew it and took measures against the exposure," Queener responded, failing to mention the bladder cancer outbreak well underway at the manufacturing site in New Jersey. Had DuPont's customers been warned? "I can't speak particularly to that point," Queener said, "but our policy is we give technical service information and caution against the certain characteristics of certain products." The inadequacy of such warnings would come into play much later, as bladder cancer surfaced and spread at Goodyear in Niagara Falls.

After the hearings had concluded, the reliably Republican-leaning and anti-regulation Chamber of Commerce sought to inflame its members with an article in its house organ, *Nation's Business*, warning that Wirtz "wants the power to shut you down" and would recruit the "hardcore unemployed" to work as safety inspectors. The misinformation crusade drew a rebuke from Senator Lee Metcalf, a Democrat from Montana, who said the Chamber and its "faithful followers have been deluging Congress with frenzied letters," many quoting the article. Metcalf entered into the record a piece in the *Catholic Standard* penned by Monsignor George Higgins, co-leader of a recently formed advocacy group, the Joint Committee on Occupational Safety and Health. Higgins decried the Chamber's "patently false attack" on Wirtz and its "obvious disdain for its own credibility. . . . Does it really have such little respect for its own membership's intelligence as to believe that such a ridiculously prejudiced article would be embraced by American businessmen, the vast majority of whom are committed to truth and fair play?" Although supporters of the administration bill were certain they held the moral high ground, the House in August floated competing legislation that darkened the mood in the Oval Office. In a memo to Califano, presidential assistant Jim Gaither lamented that the new bill was "substantially weaker" than its predecessor. It muddled the standard-setting process, gave more enforcement power to the states (which had demonstrated convincingly they weren't up to the task), and softened a provision stating employers must furnish "safe and healthful" workplaces. Distracted by the chaotic Democratic National Convention in Chicago, the war, and the impending election, Congress was never able to reconcile the two versions, and the Johnson bill died in committee. The year ended with Richard Nixon as the president-elect and hazy prospects for a law that would abate the paroxysms of death, maiming, and sickness in the nation's places of employment.

CHAPTER 4

A NEW LAW, PROMPTLY ASSAILED

IN JANUARY 1969, a shaggy-haired, nineteen-year-old student at Antioch College named Steven Wodka boarded a flight from Philadelphia to Los Angeles with his girlfriend. Inspired by a recruiter who'd come to campus the previous fall, the couple were heading west to volunteer for Cesar Chavez's United Farm Workers, then engaged in an epochal strike against grape growers to win union recognition. Raised in predominantly white, middle-class Riddlewood, a new subdivision outside Media, Pennsylvania, Wodka had been atypically single-minded for a teenager, ignoring football games and other frivolities to edit his middle school newspaper and, by his senior year in high school, become president of the Pennsylvania Federation of Temple Youth, part of a national Reform Jewish organization that opposed the Vietnam War. His father, Edward, was a chemical engineer with the Scott Paper Company, his mother, Ruth, an executive with the Elwyn Institute, a school for the mentally disabled. Ruth remembers her son as a principled and highly organized child who, unprompted, lined the shelves of his bedroom closet with shelf paper for his toys at age two or three and was infatuated with animals and rocks. "I always thought Steve was either going to be a veterinarian or the first geologist on the moon," she told me. Both Steve and his father were brilliant and wired for dogged pursuit. As he grew older, Steve developed strong anti-war sentiments and, over time, convinced Edward, a World War II veteran, that "killing is killing," Ruth said. "The father learned from the son."

Wodka attended Columbia University, Edward's alma mater, for a year, joining in the spring 1968 student takeovers of Hamilton Hall and Low Library, mounted in opposition to the university's contribution to the war effort and its construction of a de facto segregated gymnasium in a poor, Black neighborhood. Wodka was at student leader Mark Rudd's elbow as Rudd smashed a glass door to gain entry to Low, which housed the office of President Grayson Kirk. After six days, the students were ejected by plainclothes police officers swinging truncheons; Wodka avoided injury by outrunning one of the cops. He'd already grown disillusioned with Columbia by that point and was preparing to transfer to ultra-liberal Antioch, in Yellow Springs, Ohio, known for its work-study program. Impatient with Columbia's stilted "core curriculum," Wodka wanted practical experience. His destination that January day in 1969 was Delano, California, a dusty town in the southern San Joaquin Valley and the United Farm Workers' home base. He was assigned first to take photos for the union newspaper, *El Malcriado* (which, translated literally, means "The Spoiled Brat"), and went on to do research for the UFW's general counsel, Jerome "Jerry" Cohen, who was fixated on growers' bombarding of farmworkers with pesticides: chlorinated hydrocarbons, such as DDT, which embedded in fatty tissue, and organophosphates, such as Parathion, which attacked the nervous system and caused vision impairment, chest pains, convulsions—and, on occasion, death. In one especially heartbreaking episode, a three-year-old girl put her finger in a can of the organophosphate TEPP, then in her mouth, while her mother worked in a field nearby. The child vomited, lost consciousness, and died within twenty minutes. The January 15, 1969, edition of *El Malcriado* described Chavez's plans to escalate the grape boycott in response to such tragedies. "We will not tolerate the systematic poisoning of our people," Chavez had written in a letter to growers that had gone unanswered. "Even if we cannot get together on other problems, we will be damned—and we should be—if we will permit human beings to sustain permanent damage to their health from economic poisons."

The February 1 edition of the paper chronicled Cohen's unsuccessful attempts to inspect pesticide-application records in the Kern County agriculture commissioner's office—and a resulting UFW lawsuit that failed to pry loose the documents, withheld ostensibly because they contained trade secrets. ("Trade secrets may be appropriate for companies such as the Coca-Cola Company," Cohen and UFW co-founder Dolores Huerta wrote in a joint statement in November 1969, "but one drop of Coca-Cola won't kill you—and one drop of Parathion will.")

Wodka was taken with Cohen's tenacity. He went back to Antioch in April, won a $5,000 grant from the Alfred P. Sloan Foundation, took a filmmaking class, and returned to California in July to make a thirty-one-minute documentary he called *By Land, Sea and Air*. In the film, which was shown on college campuses and at meetings of environmental groups, a crop duster drenches a field in the Coachella Valley, east of Los Angeles, with an organophosphate called Dylox. Stooped workers populate the field a few hours later, while the residue is still fresh. Wodka interviews Dr. Irma West, of the California Department of Public Health, who estimates that as many as three hundred farmworkers in the state are poisoned by pesticides each year, with another thousand or so suffering less serious symptoms, such as skin or eye irritation. She speaks of an incident in Stanislaus County, in the Central Valley, in which ninety peach pickers were laid low by an oxidation byproduct of Parathion. "These chemicals," she says, "are absorbed through the skin very easily." (The concept would be relevant to Wodka's work many years hence.) Chavez goes on camera to say that workers should be tested periodically for overexposure, growers should assign registered nurses to work with the crews, and doctors should be on call. Legislation should require inspections of grapes and other produce for residue, and limits should be set. All of this would be "a good start," Chavez says, swaying in a rocking chair, but "we have to find substitutes for these very dangerous pesticides."

Wodka left California that summer deeply affected by what he'd seen. His life's course, as a defender of workers forced to choose between their health and their livelihoods, had been set. The only question was what form his advocacy would take. In early September, following up on a letter he'd sent in search of employment, Wodka met in Washington with Ralph Nader, a gangly titan in the consumer- and environmental-protection spheres and author of the best-selling 1965 exposé on the Chevrolet Corvair, *Unsafe at Any Speed*. Nader quickly sized up the young man and told him he should go to work for a union, only two of which were worth his time: the Oil, Chemical and Atomic Workers and the United Auto Workers. Nader placed a call on the spot to his ally at the OCAW, Tony Mazzocchi, and put in a good word; Wodka was hired as a research assistant, sight unseen. The union had begun in 1918 as the International Association of Oil Field, Gas Well and Refinery Workers of America—an outgrowth of a strike in the Texas oilfields the year before. It started with twenty-five members and had thirty thousand by 1921, only to experience near-extinction during the Great Depression. It rebuilt under the New Deal, changing its name to the Oil Workers International Union. That body merged with the United Gas, Coke and Chemical Workers of America in 1955 to form the OCAW. Its primary safety evangelist in the 1960s and '70s was balding, Brooklyn-born Mazzocchi, whose mother died of cancer when he was six and whose family lost its home to unpaid medical bills. Mazzocchi capitalized on the environmental movement to recruit scientists and physicians and send them into refineries and chemical plants. "He became kind of like the Pied Piper," said his biographer, Les Leopold, executive director of the Labor Institute in New York, helping shape the OCAW into a progressive force that "hit above its weight. No other union had that position in society."

Shaking off the disappointment of the failed Johnson bill the previous year, Mazzocchi had begun preparing for another run at a worker-safety law early in 1969. He organized seven regional meetings of OCAW locals—starting with one at the Holiday Inn in Kenilworth, New Jersey, on March 29—to collect evidence. Mazzocchi opened the Kenilworth meeting with a warning about the "insidious" danger all of them faced: "It's the danger of a contaminated environment, the workplace, something we don't feel, see or smell." The stories came next. A twenty-two-year-old worker at the Diamond Shamrock PVC plant in New Castle, Delaware, had been sent in to clean a tank without a spotter; his body was recovered after he'd been missing for about an hour. His corpse was blue, its tongue swollen to four or five times its normal size. He'd been asphyxiated by some undetermined poison. Company doctors who practiced in the domain of Local 8–623 in Bayonne, New Jersey, were suspected of withholding information about industrial illnesses from workers and threatening outside physicians with non-payment if they connected their patients' conditions to their jobs. Workers at Union Carbide in Bound Brook, New Jersey, believed they were exposed to high levels of mercury during electrical tests; a janitor at the plant who'd inhaled vinyl chloride fumes for twenty-five years could "hardly breathe" and was "almost blue, almost all the time." Harold Smith, with Local 8–447 and Woodbridge Chemical in Hawthorne, New Jersey, complained of inhaling mercury and chlorine. "The masks don't keep all of this out of your system," Smith said. "When I go home at night, I have roast beef and fumes. . . . If I have dessert, I can still taste some of the chemicals." His father and uncle had worked at a plant next door to his for about twenty years. Both were dead; the uncle had been consumed by throat cancer and "died like a dog." Despite these anecdotes, Mazzocchi had no illusions about the legislative battle ahead. Someone had handed him an ad from *Nation's Business* that provocatively asked, "Should Uncle Sam set your safety rules?" One OCAW lobbyist would be met with sixty from the other side. "We've got to proceed on the basis that we don't have too many friends," Mazzocchi cautioned his union brothers as he closed the meeting. They needed to show up at hearings and say, "Mr. Congressman, this is what's ailing us, these are the documented facts. We want you to do something about it, because we don't intend to tolerate, one moment longer, the murder of ourselves and our families."

Nixon was not unsympathetic to their plight. The former vice president had gotten interested in worker health and safety through his closest friend in the Eisenhower administration, Labor Secretary Jim Mitchell, a "liberal influence" in a White House dominated by businessmen, according to his *Times* obituary in 1964. Mitchell supported labor's right to organize and denounced conditions endured by migrant farmworkers and Black people. He was once called "the social conscience of the Republican Party," and the label stuck. In a speech to Congress on August 6, 1969, Nixon previewed his own bill, which would give "greater attention to health considerations, which are often difficult to perceive and which have often been overlooked. . . . Every now and then a major disaster—in a factory or an office building or a mine—will dramatize certain occupational hazards. But most such dangers are recognized under less dramatic circumstances. Often, for example, a threat to good health will build up slowly over a period of many years. To such situations, the public gives very little attention. Yet the cumulative extent of such losses is great." The bill was not as stout as Johnson's had been. Among other things, it encouraged state governments to improve their safety and health programs—an unlikely prospect, given their track record—and left the establishment of standards to a new federal board rather than the secretary of labor.

A counterproposal, closer to the Johnson bill, had already been introduced by Congressman James O'Hara and Senator Harrison Williams, a Democrat from New Jersey. Hearings in the House and Senate commenced in the fall of 1969, just as Wodka was getting acclimated to the OCAW and its front man, Mazzocchi. On November 18, 1969, Mazzocchi presented the House Select Subcommittee on Labor with another compendium of lethal and disabling illnesses suffered by the workers he represented. Wodka—not yet twenty-one, with a drooping mustache, long sideburns, and unkempt, wavy hair—joined him at the hearing table. As always, Mazzocchi's testimony was a mix of hard facts and drama. "The mad rush of science has propelled us into a strange and uncharted environment in which chemistry has taken the molecules of nature apart and reformed them into molecules which nature—man, beast and plant life—is not prepared to handle," he said. "A man lost in a tropical jungle has at least one clue to what in nature is safe to eat and what is poisonous—he can watch the monkeys and eat only what they eat. A man working in the industrial-chemical jungle of today has no guidelines on many of the fumes he breathes or the fluids which seep into his skin." He let a few of the OCAW's two hundred thousand members speak through their responses to a health and safety questionnaire completed by 130 locals and their comments at the regional conferences. (Compiling the results had been Wodka's first assignment.) F. Q. Hood from Shreveport, Louisiana, reported that exposure to exotic gases at his plant "changes the complexion of our men and bleaches their hair. Men sometimes turn blue and pass out." Albert Nist of Ashtabula, Ohio, bemoaned "the chlorine gas that's flying around" his factory. "We've had guys who have had their lungs removed . . . we've got guys who've been forced to retire because they just couldn't take it any longer."

Refinery worker John Hocking of Texas City, Texas, explained that hydrogen sulfide gas, a component of crude oil that smells like rotten eggs, had killed several members of his local. When nearby residents complained about the odor, the company injected a substance it called "Mum" into the refinery's stacks. "It's nothing but a perfume," Mazzocchi quoted Hocking as saying. National Lead, the poster child for dangerous workplaces the year before, had recorded another death from carbon monoxide. Workers at OCAW plants were still getting sick and becoming incapacitated: Stanley Wollana, Matthew Panzallela, Dan Staley, Richard Elliott, Robert Frazer, Floyd Garden—on it went, like a casualty list from Vietnam. Wodka, in his first appearance before Congress, talked about his investigation into acrylamide, an ingredient in acrylic paints that was poisoning workers at American Cyanamid plants in New Jersey and Louisiana—even those who had taken precautions. Half of one man's face was paralyzed. Another couldn't walk and was out of work. A third had a rash from the top of his head to his toes. Mazzocchi repudiated two oil-industry executives who'd bragged at a hearing several weeks earlier about the industry's excellent safety record. He allowed that there probably *were* fewer broken legs and smashed fingers among refinery workers than in years past. But the executives "said not one word in their testimony about environmental hazards. Not one word about toxic fumes and gases which kill slowly." Industry safety data was misleading because it almost never included illnesses, Mazzocchi said: "The occasional man who is crushed by heavy equipment becomes a statistic; the man who withers away with cancer, emphysema or brain damage does not."

For the rest of 1969 and most of 1970, Republicans and Democrats haggled over the particulars of the legislation. Wodka left Antioch without graduating to work for the OCAW full time. There were more hearings; at one, before the Senate in March 1970, Wodka—who'd been sent into the field by Mazzocchi to round up witnesses—introduced OCAW members from northern New Jersey, among them Emil Peter, who worked at a small chemical plant in Carteret. The plant used upward of a hundred chemicals, Peter testified, including one that could kill a man in quantities as low as 1/100th of an ounce. Over the previous eighteen months, several workers had died a year or two short of retirement age—sixty-five—having spent all of their working lives in the petrochemical industry. Herbert Ross, with Texaco in Westville, said it took Local 8–638 four years under Walsh-Healey to force corrective measures in a laboratory where one employee already had died of leukemia and others were told by their doctors, "Get out if you expect to live." The delay, Ross said, was unconscionable. At a hearing the following month, Mazzocchi presented more findings from the union's health and safety questionnaire. It showed that most fume- and dust-choked plants rarely or never were visited by state or federal safety inspectors. When inspections were conducted, the results usually weren't shared with the union. There was little difference between plants that were covered by Walsh-Healey and those that weren't. "We are living under law," Mazzocchi said, "but have never had the benefit of its protection." Even beleaguered coal miners—their lungs assaulted by silica and coal dust, in perpetual danger from explosions and cave-ins—had gotten some relief, in the form of the 1969 Federal Coal Mine Health and Safety Act.

As Congress deliberated, Ray Kline was settling into a job as a mechanic at the Goodyear plant in Niagara Falls. The money was better than it had been when he was an operator on the vinyl side, and Ray liked fixing things more than mindlessly dumping powder into bags. But there were new vexations and risks. His work took him into both Department 145 (vinyl) and Department 245 (rubber chemicals), exposing him to an even greater number of exotic substances, among them ortho-toluidine, the compound used in the antioxidant Nailax that would nearly do him in. For many years Ray knew it only as "Dominic" and had no appreciation of its potency. He encountered it while connecting and disconnecting rail tank-car lines and unplugging Nailax reactor lines, his protective equipment consisting only of safety glasses, a hard hat, and porous cotton gloves. Sometimes his lips and fingernails would turn blue while he was repairing pumps or working inside reactors or vessels called "degassers." The medical name for the condition is cyanosis, and it's usually caused by a lack of oxygen. Ray experienced it at least a dozen times. He'd enter the tanks with no respiratory protection—only a rope tied around his wrist in case he passed out and had to be extracted. He wasn't required to wear a supplied-air, full-face mask until the early 1990s, even though vessels and lines disabled for maintenance sometimes still contained solid or liquid chemicals. Ray and other mechanics would use scoops to put the gunk into cloth bags, as if it were dirt or sand instead of a malignant brew that could sicken or kill them. Ray had ortho-toluidine splashed on his hands, arms, legs, and feet with regularity; sometimes it soaked through to his skin. When he'd break a flange or take apart a pump, the liquid would spill onto the floor.

In the fall of 1970, divisions remained over the safety bill. Williams and others who favored a strong, federal role insisted that the secretary of labor be allowed to set standards, not a five-person, presidentially appointed board. Senator Jacob Javits, a Republican from New York who had carried the administration bill, argued against the "concentration of power into the hands of a single person," raising the "specter of abuse." The differences were resolved by conference committee in December. The Labor Department would set and enforce standards through a branch to be called the Occupational Safety and Health Administration, or OSHA. Where no standards existed, OSHA would have the authority to cite violations under a so-called general duty clause, which decreed that employers furnish workplaces "free from recognized hazards." A research agency, the National Institute for Occupational Safety and Health, or NIOSH, would take shape within the Department of Health, Education and Welfare, replacing the Public Health Service's overmatched Bureau of Occupational Safety and Health, or BOSH. The compromise bill easily passed both houses of Congress, and Nixon held a signing ceremony at the Labor Department just after noon on December 29, calling it "probably one of the most important pieces of legislation, from the standpoint of 55 million people who will be covered by it, ever passed by the Congress of the United States." It would deal not only with job-site butchery but also the work environment—the chemically infused netherworlds in Sayreville, Bound Brook, Hawthorne, and scores of other places Mazzocchi and his union brothers had so colorfully described. Nixon gave a nod to Jim Mitchell as he wrapped up, saying, "I am sure that he, as one of the greatest former secretaries of labor, would be very proud of the fact that this bill finally is being signed." The Occupational Safety and Health Act of 1970 was very much the progeny of Mazzocchi and, by extension, his young associate, Wodka.

The celebration didn't last. Within months, complaints were pouring in from business owners, trade groups, and their congressional allies. OSHA, which began inspecting workplaces on April 28, 1971, gained a reputation as a hairsplitting scold that cared more about the design of toilet seats and exit signs than about life-threatening hazards. Its inspectors elicited oddly visceral reactions; references to the Nazis were not uncommon. In a handwritten note to Nixon, O. William Habel, a self-described "lifelong conservative Republican and . . . Nixon booster" from Ann Arbor, Michigan, wrote that if the president had played any role in OSHA's creation, "you should be thrown out of office at once." Attached to the letter was an article in the *Grand Rapids Press* with the headline, "Federal Safety Act Will Cost Public, Industry Millions." The author of the piece warned that, in fact, costs could reach the billions and would "all be passed on to the American consumer in one way or another," starting with employers, who would "pay through the nose for violations." John P. Hughes, president of the John P. Hughes Motor Company in Lynchburg, Virginia, griped that the law was "the most potentially coercive and diabolic regulation which may put small businesses and farmers unjustly, and with very little recourse, under the iron hand of the Government bureau." In a letter to OSHA chief George Guenther, Republican congressman Dave Martin wrote, "Reports which I have had from my district in Nebraska indicate that your people are operating under this law in a high handed, dictatorial manner which is reminiscent of the days of Mr. Hitler in Germany." Nixon personally responded to Senator Clifford Hansen, a Wyoming Republican who'd also heard from angry constituents. The president reminded Hansen that the law was still fresh, "with all the attendant problems, complexities, and unanswered questions." There undoubtedly would be "individual cases in which enforcement orders seem unreasonable or inequitable," Nixon wrote, and the young agency faced a steep learning curve, especially when it came to chemicals: "Our knowledge of the effects of hazardous materials is totally inadequate. Some 8,000 substances have been identified and more are being introduced daily."

On the labor side, there was worry the law would be undermined by milquetoast regulations, government penny-pinching, and uninspired enforcement. The executive council of the Textile Workers Union of America, whose members were being asphyxiated by cotton dust, sent to the Labor Department a resolution demanding the closure of loopholes that, among other things, would allow mill owners to get advance notice of OSHA inspections, giving them time to clean up their shops. Guenther assured the union's general secretary-treasurer that the loopholes would be shut, more funds would be sought from Congress if needed, and proper implementation of the act would be the department's "highest priority." Steve Wodka, as it happened, played a part in OSHA citation No. 1, issued on May 28, 1971. A team of state and federal health investigators in February had surveyed an OCAW plant in Moundsville, West Virginia, operated by Allied Chemical Corporation since 1953, after learning several workers had suffered mercury poisoning, a crippling affliction first documented in an industrial setting some five hundred years earlier by Austrian physician Ulrich Ellenbog. The team found high levels of mercury vapor in Allied's South Plant, which it attributed to pooling of the toxic metal on the floor and working surfaces. Hearing from members that conditions had not improved, the OCAW filed an "imminent danger" complaint with OSHA on May 14, but it took five days for inspectors to arrive at the site. Once there they found "visible pools and droplets" of mercury, which can be lethal in sufficient quantities and damages the nervous, digestive, and immune systems at lower levels. Airborne concentrations were well above a voluntary industry standard that coincidentally became a legal limit the day after the citation was issued. For this infraction Allied, which recorded $55 million in earnings that year ($408 million in 2022 dollars), was accused of violating the general duty clause, fined $1,000, and given five days to correct the problem. During a hastily convened meeting in Washington, Mazzocchi, Wodka, physician Sidney Wolfe (who would go on to run the Health Research Group at Nader's Public Citizen), and Sheldon Samuels, health and safety director for the AFL-CIO's Industrial Union Department, tried to persuade Guenther and a Labor Department lawyer to shut down Allied's chlor-alkali operation—where chlorine was made by running an electrical current through brine in mercury cells—until levels of the metal were brought under control. Guenther insisted there was no imminent danger because the South Plant did not appear to be on the verge of exploding or collapsing. According to Les Leopold's biography, *The Man Who Hated Work and Loved Labor*, Mazzocchi responded, "You know, we're not that dumb-ass. We don't need to call you if something's gonna explode. We'll just run. We need you in non-explosive situations when exposures are harmful." Conditions at Allied improved incrementally, but nearly four years after the OSHA citation workers were still being exposed to mercury droplets and vapors. An OCAW consultant concluded that Allied was trying to run the plant into the ground rather than spend money to maintain and update it; it sold the facility to a company named LCP Chemicals-West Virginia in 1980, and LCP later closed the plant. The Allied episode, in Wodka's view, set an ominous precedent: to this day, OSHA has never declared any workplace health hazard to be an imminent danger. The human impacts of this hands-off policy over the past half-century can only be imagined.

CHAPTER 5

TYLER'S ASBESTOS DISASTER

THE EAST TEXAS CITY OF TYLER is known for its rose nurseries, which replaced blighted peach orchards a century ago and spawned a multimillion-dollar industry. The city calls itself, with justification, the Rose Capital of America; it and its sandy-soiled environs account for about half of the nation's rose bushes. From 1954 until 1972, Tyler hosted another enterprise it might prefer to wipe from memory: a 100,000-square-foot wood and corrugated-metal complex on the city's far northeast side built by the Union Asbestos and Rubber Company—UNARCO. The maelstrom that occurred inside—and beyond—the factory would come to symbolize all that was possible with organized labor, and all that was wrong with the nation's new regulatory framework. It would have a profound influence on Steve Wodka, still early in his tenure at the OCAW.

The Tyler plant made Unibestos pipe insulation ("so rugged it's reusable") from a type of brown, iron-rich asbestos known as amosite, mined in South Africa. It replaced a UNARCO plant in Paterson, New Jersey, that operated from 1941 to 1954 and would prove to be an incubator of disease so prolific it would draw the interest of Irving Selikoff, the nation's preeminent asbestos researcher. Pittsburgh Corning Corporation, a subsidiary of PPG Industries, bought the Tyler plant from UNARCO in 1962 and commissioned its first industrial hygiene study the following year. The study recommended that the company institute a "better housekeeping program" to hold down dust levels. It came a few months after a Walsh-Healey inspector with the Labor Department had found an "excessive amount of asbestos dust" near the sawing and scrap-grinding operations. Pittsburgh Corning commissioned a second study in 1966 to see if there had been any improvement since the first. There hadn't. Dust levels were still far above legal limits (which were themselves lenient, treating clouds of amosite as "nuisance" dust that clogged the sinuses rather than masses of needle-like fibers that easily penetrated the lungs). Ventilation was poor. The company apparently did little or nothing to fix these problems because the plant was still squalid when investigators with the federal Bureau of Occupational Safety and Health—BOSH, attached to the Department of Health, Education and Welfare—toured it in 1967 and 1970 as part of a nationwide asbestos-exposure survey, and when a Walsh-Healey inspector made his rounds in 1969. The BOSH investigators found dust levels more than ten times the limit, and fiber counts—a more accurate measure of risk—up to 1,635 times higher than what would be allowed today. This information, however, went only to Pittsburgh Corning managers; the workers and the union that represented them, the OCAW, weren't told. Nor was the state health department.

By the spring of 1971, BOSH had been replaced by the National Institute for Occupational Safety and Health—NIOSH. That summer, a young NIOSH physician in Cincinnati named William Johnson found a copy of the 1970 Pittsburgh Corning survey in a file cabinet and was struck by the sky-high asbestos fiber readings. Johnson alerted PPG's medical director, Dr. Lee Grant, who said the company had "no plans to improve ventilation in Tyler," and paradoxically "added there was a high employee turnover because of unpleasant working conditions," Johnson wrote in a memo dated August 17, 1971. Johnson called Steve Wodka in the OCAW's Washington office five weeks later and followed up by mail, sending copies of the 1967 and 1970 BOSH reports and asking if he could obtain X-rays, pulmonary function test data, and medical questionnaire results for union members at the Tyler plant. Wodka had been talking to Johnson about life-threatening conditions in another OCAW factory—Kawecki Berylco Industries in Hazleton,

Pennsylvania, a specialty-metals company that was allowing its workers' lungs to be bombarded by ultrafine beryllium particles—but hadn't been aware of the crisis in Tyler. An industrial hygiene consultant brought into the Kawecki Berylco plant by the union found extremely high levels of beryllium dust, as did OSHA, which fined the company only $600. Kawecki Berylco's response was to requisition new respirators for the workers—small comfort for men like forty-four-year-old Robert Ferdinand, who "had to take oxygen four times a day," the *New York Times* reported, and whose coughing spells had become "so violent" that his doctor feared he'd develop a hernia. Kawecki Berylco's plan didn't fly with Mazzocchi and Wodka, who insisted that the dust be controlled at its source and the OCAW local have the right to keep monitoring the air. The company made concessions, but not enough for the union's liking, leading to a five-month strike in 1973—possibly the first in US history "where the primary issue was worker health and safety," *Newsday* reported. The strike ended when Kawecki Berylco agreed to pay for annual employee physicals, independent industrial-hygiene surveys, and other measures that led to vastly improved conditions. It was a rare good-news story at a time of gloom in the workplace, an example of how labor and management could find common ground without fighting interminably or deferring to an anemic OSHA.

Johnson's call to Wodka about Tyler sparked a burst of activity over the next four months. In late October of 1971, Johnson and two NIOSH colleagues, physician Richard Spiegel and epidemiologist Richard Lemen, went to Tyler to see things for themselves. It was worse than they'd imagined. Inside the Pittsburgh Corning plant, seventy-four employees wheezed and coughed amid dust that collected on floors, ceilings, and drinking fountains. The dust had been pushed into piles near the machines by workers with brooms, allowing it to be re-suspended within the factory's two buildings when it was disturbed. The indoor air was opaque and stifling; workers recently had been given respirators to compensate, but the masks were not properly fitted, worn, or maintained and, therefore, did little good. The NIOSH investigators tried to take measurements in the ventilation system, but it was so clogged with asbestos that they had to use broom handles to pull the stuff out of the vents in clumps, as if it were cotton. The lunchroom was located only forty to fifty feet from the dustiest part of the plant, raising the strong possibility that the workers were ingesting amosite fibers along with their baloney sandwiches.

Seven of eighteen workers who had been employed ten or more years showed signs of asbestosis, an inflammation and scarring of the lungs usually caused by very high exposures to asbestos fibers for an extended period. Several workers with less than five years at the plant already showed reduced pulmonary function. The hazards weren't confined to the plant: a field next door was used as a dump for asbestos waste. "At the present time, no provisions have been made to bury this material," the NIOSH investigators wrote in their trip report. In fact, there were two more dumps—one about the size of a football field—near a city sewage treatment plant. Pittsburgh Corning, moreover, had decided it was good business to sell the burlap bags in which raw asbestos had been stored to local nurseries, which used them as wraps for rose bushes and other shrubs. "In conclusion," the investigators wrote, "an extremely serious and critical occupational health situation exists at this plant. Immediate corrective action is necessary to reduce asbestos exposures to conform to existing standards." PPG's response was to announce that it would close the plant "for economic reasons" by January 31, 1972.

In a letter on November 16, Johnson, the NIOSH physician, alerted the Texas State Department of Health to "an extremely serious and critical occupational health situation" in Tyler. This, in turn, led to an OSHA inspection late in the month; the inspectors found, to no one's surprise, excessive asbestos dust levels and ineffective ventilation, problems that could be fixed with "major modifications" to the plant. The point seemed moot, given Pittsburgh Corning's decision to shut down a few months hence, but when OSHA proposed a $210 fine on December 16, Mazzocchi was apoplectic. He went public at the annual meeting of the American Association for the Advancement of Science in Philadelphia ten days later. "Mazzocchi zeroed in on Tyler as the quintessential example of the ignorance, neglect, and subterfuge that characterized government policy toward industry hygiene," journalist Paul Brodeur wrote in a 1973 *New Yorker* article. "Mazzocchi believed that the [Nixon] Administration's people were hoping that if they slapped Pittsburgh Corning lightly on the wrist with some nonserious violations the whole affair would blow over. . . . It became apparent in December of 1971 that the Tyler affair was not going to blow over. That it did not, as so many other occupational-health scandals had, was largely the result of the efforts of Mazzocchi, who was determined to make it a cause célèbre." Meanwhile, Grant, the PPG medical director, responded by writing a memo for his files seeking to absolve himself of responsibility for the Tyler catastrophe and clear up a "misunderstanding" between him and Johnson. Grant claimed he had a "long-standing interest in the health of the Tyler employees," and blamed a four-year delay in worker medical examinations on BOSH and personnel shortages at the East Texas Chest Hospital. In fact, Grant was less benevolent than he made himself out to be, sharing the results of the exams not with the workers but with plant manager Charles Van Horne, hardly an unbiased medical authority. As NIOSH epidemiologist Lemen remembers it, Van Horne cavalierly explained to workers who'd been diagnosed with pleural calcification—a classic marker of asbestos exposure—that the condition was caused by the consumption of too much milk.

Wodka made his first visit to Tyler early in 1972, arriving late in the day on January 6. Although the Pittsburgh Corning plant was only weeks from closing, he was determined to see the company take basic measures it had disregarded: making sure the workers had respirators, didn't expose their families to asbestos by taking their dirty work clothes home, and were followed medically. The next morning, Wodka met with Grant (who identified himself without irony as president of the American College of Preventive Medicine), Lemen, local OCAW president Herman Yandle, and Dr. George Hurst of the East Texas Chest Hospital. After the meeting, Wodka was given a tour of the plant he'd heard so much about. He saw what he would later describe as "a mountain of raw asbestos fiber in the middle of the plant, dust blowing around, dust everywhere." Wodka, Lemen, and Yandle drove next to an open field near State Highway 155 and Interstate 20, where weeds poked through piles of asbestos. Wodka took photos. From there it was on to Eikner Nurseries, which had bought countless burlap bags from

Pittsburgh Corning over the years and still had, by Lemen's estimate, fifteen hundred to two thousand of them stacked in a barn. The visitors saw a Latina worker shake the dust from one of the bags and use it as a lap cover while she grafted plants on a chilly day. Lemen took a bag back to the NIOSH lab in Cincinnati for testing; "extremely high concentrations" of asbestos fibers were found, according to a report by the agency's Environmental Investigations Branch.

The week after Wodka's trip to Tyler, OSHA—apparently chastened by the fuss Mazzocchi had raised over the $210 fine—sent an inspector back to the Pittsburgh Corning plant. The inspector found no sign of progress since November and hit the company with $6,990 in "failure to abate" penalties. On February 3, Pittsburgh Corning ended production in Tyler; five days later, OCAW president A. F. Grospiron sent telegrams to OSHA and the Environmental Protection Agency, pleading with them to no avail not to drop their investigations into the company's behavior inside and outside the plant, which had employed some 1,500 people in its eighteen years of operation. In fact, the damage had already been done. In a September 1973 report, three NIOSH investigators predicted "the most serious consequences of the asbestos exposures resulting from this facility . . . may be many years in coming," given that the latency period for lung cancer and mesothelioma was twenty to thirty years. There was evidence such cancers could materialize "after only a casual exposure," the investigators wrote, adding that "mortality studies have hinted at increased cancer rates at other sites, such as the gastro-intestinal tract." Decades later Hurst and three other researchers would find significant excesses of mesothelioma and cancers of the lung, trachea, and bronchus among 1,130 former Pittsburgh Corning workers. Even exposure durations of less than six months produced heightened risk of death from respiratory cancers. Elevations of gastrointestinal cancers—mostly of the colon—also were noted, as the NIOSH investigators had forecast. No one knows what happened to the nursery workers who handled the dusty burlap bags, or people who lived near or otherwise encountered the asbestos dump sites.

Pittsburgh Corning and UNARCO paid, in a very modest sense, for their misdeeds in Tyler, contributing $8.1 million and $1 million, respectively, toward a $20 million settlement—finalized in 1978—of lawsuits brought by 445 former workers. Asbestos supplier Cape Industries Limited kicked in $5.2 million, the US government $5.7 million (the latter for withholding health information from the workers). In a 1974 letter to the *Journal of Occupational Medicine*, Johnson, the NIOSH doctor who first reached out to Wodka about the Tyler plant, decried the Public Health Service's role in the disaster. The agency's asbestos field study, begun by BOSH in the late 1960s, operated under an indefensible veil of secrecy, Johnson wrote. Government investigators "were not provided respirators, even in the dustiest of the asbestos plants, because workers might know they were breathing a toxic dust." In most cases survey results were shared, via prior agreement, only with the companies. The initiative, Johnson argued, "contains more elements of an atrocity than the Tuskegee Study of 'Untreated Syphilis in the Male Negro.'" But few in his profession, likely hoping to avoid "the embarrassment of the medical/industrial complex," wanted to discuss it. Wodka's takeaway from Tyler was more sanguine. The OCAW local that represented the Pittsburgh Corning workers had voted for more government inspections of the plant, knowing that such inspections would likely expedite its closure—and keep it shuttered permanently. The local's posture signified what Mazzocchi called a union's "moral imperative"—putting its members' health above its institutional need to collect dues—and was a stark contrast to what would happen at Goodyear in Niagara Falls a half-century later.

CHAPTER 6

VINYL

O N JANUARY 22, 1974, B.F. Goodrich notified NIOSH that four of its polyvinyl chloride workers in Louisville had developed angiosarcoma of the liver, an extremely rare cancer. John Creech, the Goodrich plant physician who had first reported the hand condition acroosteolysis among reactor cleaners a decade earlier, was again the sentinel. Creech had noticed that more than a few workers in Louisville were returning from sick leave with liver abnormalities. Some were jaundiced and vomiting blood. One of his patients had died almost two years earlier of angiosarcoma, but he'd written it off as an anomaly. Then two more workers, under the care of different doctors, were felled by the disease in March and December of 1973. Creech and Goodrich environmental health director Maurice Johnson concluded that the deaths almost certainly were work-related and notified corporate executives in Akron, who notified NIOSH. A fourth case was confirmed not long afterward—a plant worker who'd died in 1968 of what mistakenly had been classified as a primary liver tumor. NIOSH's parent, the Centers for Disease Control, reported that only twenty-five cases of angiosarcoma occurred each year in the entire United States; four from one plant was "a most unusual event" that suggested a workplace carcinogen was responsible. It seemed "distinctly possible" the problem was industry wide, the CDC said.

The news set in motion a chain of events that would change the course of the PVC industry and create an archetype for decisive regulatory action. In February 1974, two more cases of angiosarcoma were found in Louisville, bringing the total at the Goodrich plant to six. Only one of the victims was still alive. Leaders of OCAW Local 8–277 in Niagara Falls, worried about the health of their own members, asked Irving Selikoff's team from Mount Sinai to examine workers from the Goodyear plant. The scene at the union hall was reminiscent of "the field hospital in 'M*A*S*H,'" the New York Times reported. "In the union's second-floor offices, examining beds were fashioned from long steel tables covered with blankets and paper sheets. A local motel lent pillow cases and ice buckets to hold more than 2,000 tubes of blood." One confirmed case of angiosarcoma had been found, as had a suspected case.

In April, OSHA adopted an emergency temporary standard for vinyl chloride of 50 parts per million, ten times more stringent than the exposure limit at the time. That summer, OSHA held hearings at which manufacturers groundlessly complained they'd be driven out of business if the limit was too strict. The agency adopted a permanent standard of 1 ppm in October; the Society of the Plastics Industry, a trade group, challenged it in court and lost. For OSHA, it was a remarkable achievement that would not be replicated. It had identified a hazard and brought it to heel within a single calendar year. Twenty-seven thousand workers with known exposures and 2.2 million with potential exposures to a virulent carcinogen were the beneficiaries. PVC makers did not shut down en masse, as they'd predicted. They adjusted to the new vinyl chloride limit and continued to turn out the profitable white PVC resin; two years after the standard took effect, Chemical Week reported that producers were revamping operations and building units to meet growing market demand.

In 1975, a decade after the two deformed Kline children were buried, Peter Infante, a frizzy-haired young epidemiologist with the Ohio Department of Health, published a paper summarizing research he had done in three centers of PVC manufacturing in that state: Painesville, Ashtabula, and Avon Lake. Infante, who went on to work for NIOSH and OSHA, found statistically significant elevations of central nervous system defects, including anencephaly, spina bifida, and hydrocephalus, in babies born in those cities. Infante did not incriminate vinyl chloride specifically, but the publication of his study was enough to cause heart-burn at the CDC. In an October 1975 letter from J. William Flynt, chief of the agency's Birth Defects Branch, to Joseph K. Wagoner, NIOSH's director of field studies and clinical investigations, Flynt complained that he had just read quotations about the study from Wagoner in a wire-service story and learned that journalist Dan Rather had discussed it during a CBS newscast. Flynt—who had called Infante when he was still in Ohio to voice similar misgivings—wrote that there was "little basis in fact for continuing to state that communities having vinyl chloride plants are at increased risk of defects. Such statements only cause needless concern and anxiety among these and other communities." In a memo to NIOSH director Jack Finklea a week later, Wagoner and Infante wrote, "Not only are we concerned that birth defects may be related to VC [vinyl chloride] exposure, but also that VC exposure may be related to excessive fetal loss." The findings in Ohio were "highly consistent with experimental studies conducted by Dow Chemical Company which show VC to be teratogenic [capable of causing malformations] in laboratory animals," Wagoner and Infante wrote. Vinyl chloride, they added, also might be mutagenic—capable of altering genetic material to ill effect.

Flynt and two associates from the CDC's Chronic Diseases Division followed up with their own study of Kanawha County, West Virginia, a hub of PVC and other chemical manufacturing that had shown an excess of central nervous system defects in the early 1970s. While noting that a number of these cases were clustered to the northeast of an unidentified PVC plant (Union Carbide's)—i.e., downwind from it much of the time—the investigators reported that defect rates were on the decline, and vinyl chloride was but one of more than a hundred toxic compounds spewed by the area's factories. In short, there was nothing to worry about. But Infante thought there was persuasive evidence of a connection. As the *New York Times* reported:

In addition to birth defects among the general population living near vinyl chloride plants, Dr. Infante has studied the results of pregnancies among the wives of vinyl chloride workers. In a report to be published in the British journal, The Lancet, Dr. Infante states that he found a two to three-fold increase in the rate of stillbirths and miscarriages among the wives of vinyl chloride workers, when compared to a control group whose husbands were not occupationally exposed to vinyl chloride.

In the early and mid-1960s, Ray Kline would have inhaled huge amounts of vinyl chloride while cleaning PVC reactors. He brought home his dirty work clothes for Dottie to wash once or twice a week. At the time, PVC resin contained lots of residual vinyl chloride monomer, exposing Dottie while she was pregnant. She also may have been affected by offsite vinyl chloride emissions from Goodyear, which were estimated by the EPA at four tons per year in 1975 and probably were much higher during Dottie's pregnancies. Maybe a combination of exposures laid waste to John and Dona Kline while they were in the womb.

The PVC industry, as it happens, suspected something was awry with vinyl chloride long before Goodrich went public in 1974. Dow Chemical had begun discreetly testing the chemical on rats, rabbits, guinea pigs, and dogs at its Biochemical Research Laboratory in Midland, Michigan, in 1958, and the results—shared with Goodyear and other companies that belonged to the Manufacturing Chemists' Association, or MCA—were troubling. Adverse liver effects had been seen in animals that had inhaled only 100 parts per million, a fraction of the concentration to which many workers were exposed. In a letter to a high-ranking Goodrich toxicologist and industrial hygienist on May 12, 1959, one of the Dow scientists, V. K. Rowe, outlined the experimental findings and concluded that vinyl chloride could produce "rather appreciable injury" among workers routinely exposed to 500 parts per million, then the voluntary workplace standard. Rowe ended his letter with a plea: "This opinion is not ready for dissemination yet and I would appreciate it if you would hold it in confidence but use it as you see fit in your own operations." In an intra-company memo on November 24, 1959, Union Carbide toxicologist Henry F. Smyth Jr. wrote that even 100 parts per million "produced organ weight changes and gross pathology. . . . Vinyl chloride monomer is more toxic than has been believed." Before Dow began its animal testing, there had been few investigations of the chemical, though acute effects had been seen in guinea pigs as early as 1930, and liver damage had been found in fifteen of forty-five vinyl chloride workers examined as part of a 1949 study in the Soviet Union. The Dow researchers published their data in the *American Industrial Hygiene Association Journal* in 1961, recommending a vinyl chloride exposure limit of 50 parts per million. No one outside of Dow listened. "The industry goofed up," Theodore Torkelson, a toxicologist who helped conduct the experiments, admitted nearly four decades later.

In 1969, Dr. Pier Luigi Viola, an industrial physician for the Belgian chemical firm Solvay, attempted to produce acroosteolysis in rats exposed to vinyl chloride gas in a Rome laboratory. Instead, he produced cancer, albeit at very high concentrations. American manufacturers grasped the significance of this unexpected finding. "Since this report appears to be so convincing," Dr. George Roush, medical director of the Ethyl Corp. in New Orleans, wrote to a colleague on June 24, 1970, "I suppose that we must consider vinyl chloride as a carcinogen." By the end of 1972, Dr. Cesare Maltoni, an Italian oncologist, had seen angiosarcoma of the liver in rats exposed to as little as 250 parts per million, a fraction of what reactor cleaners like Ray Kline would have inhaled. The news traveled quickly across the Atlantic. An Allied Chemical memorandum called Maltoni's results "disconcerting," and a Uniroyal memo concluded, "The work in Europe indicates we have a problem which cannot be ignored." Members of the MCA, including Goodrich and Goodyear, signed a secrecy agreement with their European counterparts, promising not to disclose Maltoni's findings. It was an act of hubris, driven by fear of liability and regulation.

Years later, Dr. Marcus Key, the first director of NIOSH, testified in a deposition that the agency was deceived by members of the MCA's vinyl chloride task group. At a meeting with him and four agency staff members in July 1973, the industry men made no mention of Maltoni's consequential studies, Key testified. He said he didn't learn of them until the Goodrich angiosarcoma cluster was revealed six months later, a delay that kept NIOSH from issuing a timely alert to employers and workers in the PVC industry and notifying OSHA, the regulator. Members of the MCA task group disputed Key's account, claiming they'd told the

NIOSH officials that Maltoni had found cancer in animals at 250 ppm. No one knows with certainty what happened at the meeting. But the PVC industry would later be accused of engaging in a conspiracy, an allegation supported by memos, letters, and minutes exhumed in litigation.

A NIOSH alert might have saved William R. Smith, one of Harry Weist's coworkers in Department 145 at Goodyear in Niagara Falls, who died of brain cancer in September 1993. Smith began working at the plant as a production operator in November 1973, when he was twenty. He cleaned PVC reactors, unloaded liquid vinyl chloride from tank cars, and performed other duties that put him squarely in the high-exposure category. All told, he spent more than thirteen years in Department 145. The first sign of trouble came in October 1992, when Smith's neck grew sore and stiff—though he'd suffered recurring headaches before that. By Thanksgiving, his neck pain had worsened. The base of his skull, behind his right ear, was sensitive to the touch. He complained of "numbness in his fingers and leg, just like pins and needles" after taking a hot shower at the plant, his wife, Holly, testified in a deposition. Just after Christmas, Bill became nauseated and assumed he'd caught the flu. He ate a bowl of cereal and "it came right up," Holly said. He began slurring words. A trip to the emergency room at Niagara Falls Memorial Hospital was unenlightening: a CAT scan and an EKG were negative for any worrisome pathology.

By the beginning of 1993, Bill's nausea had intensified and he'd all but ceased eating and drinking. He was admitted to Memorial, and a neurologist raised the possibility of a brain tumor—though none was found. He was transferred to Sisters of Charity Hospital in Buffalo, where an emergency MRI revealed an enlarged medulla oblongata, a critical, information-relaying part of the brainstem. Surgery exhumed a Grade 1 tumor, which the doctors considered treatable and gave Holly reason for hope. That hope was misplaced: Bill spent the next eight months in the hospital, unable to speak and hold down even soft food, embarrassed to see his seven-year-old daughter, Ashley. He was forty years old when he died. Wodka represented Holly Smith in a lawsuit against three vinyl chloride suppliers to Goodyear that was settled out of court. It was Wodka who deposed Key and got him to tell the story of his meeting with the PVC industry representatives in July 1973. Key was still indignant. He underscored in his testimony that the rapid action taken by the federal government in early 1974—a NIOSH memo urging OSHA to adopt an emergency temporary standard for vinyl chloride, which it did in a matter of weeks—would have occurred sooner had he known what the industry men knew about Maltoni's work as early as January 1973. "I would have transmitted this information to OSHA," Key said. "I would have put out a public alert on the potential problem, and I would have begun to look for human cases of cancer among those exposed to vinyl chloride, heavily exposed for a number of years." Some or all of this, presumably, would have taken place before Bill Smith began climbing into PVC reactors at Goodyear.

CHAPTER 7

HARRY BREAKS FREE

IN JUNE OF 1974, a week after graduating from Niagara Wheatfield High School in Sanborn, New York, Harry Weist left town. Harry felt the need to get away from his dysfunctional family—a kind but workaholic father, Daniel, and a volatile mother, Jeanie, who was likely bipolar. He had battled Jeanie throughout his teenage years, resisting her attempts to micromanage his life and retreating to friends' houses when things got too tense. A wiry, kinetic athlete who organized neighborhood football and baseball games and wrestled all four years in high school, Harry once defied his mother by riding his ten-speed bicycle 129 miles to Kane, Pennsylvania—where his maternal grandparents lived, and where he'd spent summers since ninth grade—in a day. Now he was free for good, again headed for Kane, where jobs were plentiful and friends from previous visits awaited.

Harry was born in Niagara Falls on March 12, 1956, and moved to rural Sanborn, ten miles to the northeast, when he was seven. He had a younger sister, Linda, born in 1958, and a brother, Barry, born in 1962. His father, from the stubby mountains of south-central Pennsylvania, held several factory jobs before settling in at Harrison Radiator, a division of General Motors in Lockport, New York. He repaired radiators for thirty-plus years and served in the 107th Air National Guard, which sent him to Vietnam in the late 1960s, though he didn't see combat. He was well liked but tyrannized by Jeanie, who, Harry says, "hated everybody and fought with everybody." The two separated and considered divorcing in the early 1970s but reconciled; Daniel later told Harry the reconciliation was the biggest mistake he'd ever made. To avoid his wife's frequent tirades, Daniel busied himself with work.

When Harry got to Kane after graduation, the little town was thriving. It was a manufacturing hub, with factories that made broom handles, carbon resistors, and—at Kane Manufacturing, where Harry landed a job—fortified window screens for prisons. Harry lived at the home of Fred Johnson, a friend of the same age whose mother, Dody, a sweet-natured widow with four other children in the house, left them blissfully alone. The boys spent much of their free time drinking in bars and smoking marijuana. When someone from Kane Manufacturing would come by looking for Harry, Dody would cover for him as he slept in a closet. Money was tight; when Harry or Fred shot a deer in the Allegheny National Forest, Dody would skin it and cook the meat for the household, wasting nothing. The game was supplemented with bread, pastries, and milk the boys stole from a supermarket.

As Harry was enjoying his freedom in Kane, Steve Wodka, Tony Mazzocchi's right-hand man at the OCAW, became entangled in a saga that would make national news and inspire a Hollywood film starring Meryl Streep. Wodka was the point of contact for union member Karen Silkwood, who worked at a Kerr-McGee nuclear fuel-rod plant near Cimarron City, Oklahoma. Silkwood was among eighty-nine Kerr-McGee workers exposed to the radioactive metal plutonium at the plant in excess of allowable limits. She later alleged that there were systemic quality-control problems and was ready to go public. Silkwood died in a one-car accident on November 13, 1974, as she was on her way to meet Wodka and *New York Times* reporter David Burnham at a Holiday Inn in Oklahoma City. Earlier that month, Silkwood's hands, nasal passages, and apartment had been tainted with plutonium, a pollen-sized speck of which could cause cancer if ingested. The incident triggered decontamination protocols and raised suspicions within the OCAW leadership that Silkwood had been deliberately poisoned.

A rookie trooper with the Oklahoma Highway Patrol who investigated the fatal wreck characterized it as "a classic one car sleeping driver type accident." Neither Mazzocchi nor Wodka bought the official explanation, noting among other things that the accusatory documents Silkwood was supposed to bring to the hotel were not found in her smashed Honda Civic. Both believed she'd been forced off the road by someone trying to silence her. For Wodka, Silkwood's violent demise was a revelation: he'd been prepared for her to be fired for her activism but hadn't envisioned anything worse than that. No one was arrested in connection with Silkwood's death. An FBI investigation was no more thorough than the Highway Patrol's. Kerr-McGee was never penalized by the US Atomic Energy Commission, even though one prominent physicist told Congress the Cimarron City plant was the sloppiest facility of its type he'd ever seen. And the workers who had been exposed to plutonium were never followed to see if they developed cancer or merited compensation. Yet Wodka didn't view this chapter in his life as a disappointment. The month before Silkwood died, Kerr-McGee workers voted against decertifying the OCAW as their bargaining unit. The vote vindicated Silkwood's and the union's efforts to correct safety hazards at the plant and keep workers from being contaminated with plutonium. A jury verdict in a lawsuit brought against the company by Silkwood's father—$505,000 in actual damages and $10 million in punitive damages—offered validation of Silkwood's allegations. And Silkwood herself—one of the original whistleblowers—became a role model for women in the labor movement.

Six weeks or so after Silkwood met her end, eighteen-year-old Harry Weist returned home to Sanborn from Pennsylvania, having been let go by Kane Manufacturing. He shared, with a high school friend, an apartment on Falls Street in downtown Niagara Falls and went to work at a plastics factory, which closed after a month. Seeing few other options, he enlisted in the Air Force in January 1975 and reported for basic training at Lackland Air Force Base in San Antonio two months later. He became a military policeman, staying at Lackland until summer, when he was transferred to Eglin Air Force Base in the Florida Panhandle. At Eglin he was caught smoking marijuana on several occasions, was ordered to enter a drug-treatment program, got caught again, and was transferred to the supply squadron. He delivered aircraft parts on the base and found that he enjoyed it. It played to one of his strengths: an ability to talk to almost anyone and get people to like him. Caught smoking dope yet again, Harry was spared a dishonorable discharge by a commander who appreciated his work ethic and wrote a letter of support. Harry left the service in January 1977 and returned to western New York just before a record-setting blizzard buried the region in up to 100 inches of snow. He collected unemployment for a time, cleared snow for the Town of Niagara under a federal worker-training program, and pondered his next move.

CHAPTER 8

EULA

BY THE MID-1970S, Steve Wodka had settled into a pattern. His experiences with Allied, Pittsburgh Corning, and Kerr-McGee had shown him the worst of corporate America—its capacity for duplicity, its callousness toward the workers who generated its profits—and revealed the impotence of America's regulatory system. Wodka, not yet thirty but hardened by more than five years of conflict, would visit an OCAW plant in eastern Pennsylvania or southeastern Texas, hear horror stories from the frightened employees, file a complaint with OSHA, grouse when the agency proposed what was often an infuriatingly low penalty, and help the union's lawyers respond to the inevitable court challenges from the companies that had been cited. (Image-conscious corporations did not—and still don't—like to admit fault and often fought even the most minor citations on principle.) Later in the decade, Wodka would see tangible results from his efforts after Eula Bingham, a no-nonsense Kentuckian, was tapped to lead OSHA and set industry back on its heels. It would not last.

The workplace health crisis extended beyond the OCAW's factories. In a front-page story on April 18, 1976, the *Akron Beacon Journal* reported that at least seven workers at the Goodyear Aerospace complex in that city had died of leukemia over a ten-year period; all had worked with the solvent benzene in Plant C, where Pliofilm, the popular packaging material, was made from 1940 until 1965. (The product was pulled from the market in 1978.) The newspaper quoted Goodyear's vice president for manufacturing, F. Vincent Prus, as saying that seven deaths in such a short time was "suspicious as hell." On May 13, the *Beacon Journal* said the number of dead from Plant C had grown to eleven. A research team from the University of North Carolina by this time had found six leukemia deaths at a Firestone synthetic rubber plant in Akron. The revelations prompted an urgent letter from the United Rubber Workers' international president, Peter Bommarito, to Labor Secretary William J. Usery Jr. "The current OSHA standard for benzene is ridiculous," Bommarito wrote. "There already appears to have been entirely too much indecision and procrastination regarding worker exposure to chemicals that have shortened the lives of my membership and produced early retirement due to occupational disease." Usery responded that a new standard was under consideration (though it wouldn't be issued in temporary form until 1977, in permanent form until 1978, and wouldn't take effect until 1987).

Fear of toxic chemicals, whose effects were becoming evident, suffused organized labor. The United Steelworkers brooded over, among many other things, emissions from sooty coke ovens, where coal was cooked to make pure carbon used in iron and steel. The vapors that came out of coke batteries were as awful as could be imagined, a boiling stew of hydrocarbons, with names like benzo(a)pyrene, tied to cancers of the lung, trachea, kidney, prostate, and other sites. At a Labor Department hearing on December 18, 1975, the Steelworkers argued for a tightening of the coke-oven standard and offered as a witness member Roosevelt Johns, who had worked at Republic Steel's coke plant in Warren, Ohio, for twenty-five years, suffered two heart attacks, and developed a tumor on his left lung. Johns was forced to retire on a disability pension at age forty-five and, during a 151-day stretch between November 1, 1974, and March 31, 1975, spent sixty-nine days in the hospital. Six union panelists at the continuation of the hearing on December 19, all of whom had worked at Bethlehem Steel's Sparrows Point coke plant near Baltimore, were in various states of disrepair: two had cancer of the larynx, one a tumor in his neck, the other three lung cancer or another lung disorder. "These courageous individuals will not be helped by a strong, comprehensive OSHA Standard," the Steelworkers wrote in a post-hearing brief. "However, any failure on the part of the [Labor] Secretary to promulgate such a Standard will be a slap in the face of their courageous sacrifice in appearing at the hearings on behalf of their working brothers and sisters." Company executives who had attended the proceedings to that point were conspicuously absent during the sick men's testimony.

The Steelworkers got their standard in October 1976. It was the first one that required employers to curb emissions with engineering controls—enclosing the coke-making process, in this case—rather than rely on cheaper half measures. But it barely made a dent in the bigger emergency. At a conference in England that May, Ralph Nader and the head of his Health Research Group, Sidney Wolfe, noted that OSHA had inspected only 4 percent of the nation's workplaces during its first four years and said that "the probability of federal inspection in any year remains painfully low." The problem was compounded by the agency's "apathetic enforcement policies." Fewer than 2 percent of OSHA citations, Nader and Wolfe found, had been handed out for serious violations, which carried bigger fines because they reflected hazards that threatened life and limb. Things were especially dismal on the health side of the equation: although numerical limits for many common chemicals were in place, many of them were weak, and OSHA had "made it clear that employers may exceed these established levels by almost three times" and still be cited for only a minor violation. The average penalty for all violations in 1975 was $25.69. A rare serious violation brought an average fine of $607, a nonserious one $13.26—the cost of two long-playing record albums or ten gallons of milk. This was hardly a deterrent to negligent behavior, given that a pattern of lesser infractions, left unaddressed, often culminated in accidents or needless exposures.

In a letter to Labor Secretary Usery on March 5, 1976, AFL-CIO president George Meany wrote of his dismay in reading a *New York Times* article claiming OSHA was considering the postponement of health standards until after the presidential election. "This indicates a repeat of the 1972 attempt by the Nixon campaign to barter the health and safety of American workers for campaign contributions to an incumbent

Republican President," Meany wrote, referring to a pledge by OSHA head George Guenther not to issue any controversial standards until Nixon had secured a second term. Usery responded that he, too, had been upset by the *Times* piece but assured Meany there was "no truth in the implications of political maneuvering set forth in the article. . . . The time required to deal with the complex scientific and technical requirements in developing safety and health standards is the result of many factors but not politics." Meany was mollified—for a while. On September 27, however, he again wrote to Usery, this time to complain about a Ford administration policy, carried over from the Nixon years, requiring that each standard be accompanied by an "inflationary impact statement"—an analysis to ensure that the benefits of a regulation outweighed its costs. Such statements were "a means of providing business management with a sounding board to set forth exaggerated claims" about the costs of complying with new rules, Meany wrote. It was a "macabre charade. . . . This device is sucking the life out of an orderly and expeditious standard development process. It has resulted not only in insufferable delays, but perverted the Act to that of a protective operation to shield employers from its legitimate consequences."

All of this came against the backdrop of what Samuel Epstein, a physician on the faculty at Case Western Reserve in Cleveland, called a "major epidemic of cancer" that was killing one in five Americans and costing, at a minimum, $15 billion a year. Experts believed 70 to 90 percent of cancers were environmentally induced and therefore preventable, Epstein wrote in a paper he presented at a conference in Washington in early 1976. He pleaded for a government chemical-testing regimen that did not then exist and in many ways still doesn't. Too often, he wrote, industry scientists were being pressured to develop and interpret data on carcinogenesis that would be "consistent with short-term marketing interests." As a result, certain "mythologies" had set in. One was that animal carcinogens were somehow less dangerous than proven human ones and should be regulated as such "until their carcinogenic effects can be validated by human experience." Epstein shot this down emphatically, writing that, with the possible exception of a certain form of arsenic, "all chemicals known to produce cancer in man . . . also produce cancer in experimental animals, generally in rodents." Vinyl chloride was a case in point. Another myth was that "safe" levels of carcinogens could be determined. This, Epstein wrote, was an example of industry bending the narrative to its commercial needs. There *were* no safe levels. Yet OSHA had stiffened the rules on only sixteen carcinogens, out of thousands, since its founding in 1971.

OSHA was taking heavy flak from business as well as labor. An unbylined, special report in *Factory* magazine in August 1976 captured the sour mood and foreshadowed agitprop that would later be deployed by right-leaning politicians on any number of fronts. Americans' "collective disgust" with OSHA "pales in comparison to what the nation's founding fathers might feel if they were still among us," the piece said. "When those forerunners of the federal apparatus framed the Constitution two centuries ago, they certainly didn't dream that one day a fourth branch of government would evolve—today's sprawling, expansive, heavy-handed and non-elected regulatory agencies. A bureaucracy that mushrooms like a cancer. Agencies that go largely uncontrolled and more often than not fail to accomplish what they set out to do. The Occupational Safety and Health Administration is perhaps the most important, the most powerful, and yet the most obvious failure among them." It had put "trivia ahead of safety" and written "endless rules . . . many of them unintelligible and downright idiotic." President Gerald Ford, running for re-election, joined the chorus, telling a crowd during a campaign stop in New Hampshire, "I'm sure you'd all like to see OSHA dumped in the ocean." Ford's overture to aggrieved businesspeople became moot when he lost to Jimmy Carter. In a parting report, Morton Corn, who led OSHA under Ford, defended his tenure. The agency had made a concerted effort to hire health and safety professionals as inspectors to combat the perception that they were all pencil-wielding simpletons. It was placing new emphasis on health, as opposed to safety alone, and aggressively recruiting industrial hygienists. "With the resources currently planned for the Health Standards Directorate of OSHA," Corn wrote, "I estimate that a productivity rate of 15 to 20 health standards promulgated per year is a noble ambition in 1978 or 1979." It was a goal that would not be met.

Soon after he was inaugurated in January 1977, Carter went to Labor Department headquarters on Constitution Avenue with a message of reassurance: the Occupational Safety and Health Act had great potential to improve workers' lives and should be rigorously enforced. Carter promised to nominate a woman to lead OSHA and said he already had someone in mind: Eula Bingham, an environmental health researcher who specialized in carcinogens. Bingham had been recommended for the job by women's groups, unions, and other scientists. An only child, she grew up on a farm near Burlington, Kentucky, and earned degrees in chemistry and biology from Eastern Kentucky University. She did her master's and doctoral work at the University of Cincinnati and was hired in 1961 as a researcher at its College of Medicine, where she conducted cancer studies underwritten by oil companies, painting chemicals on mice to see if she could induce tumors. At a seminar one Saturday morning early in her career, she brashly asked the presenter, an industry scientist, whether he thought benzene caused cancer. He responded without hesitation that he did; that's why she and others in the lab weren't allowed to dilute samples with it. That didn't keep the companies from fighting tooth and nail against benzene regulation, a strategy that surely cost lives.

Bingham studied bladder cancer in dye workers and histoplasmosis, a lung infection caused by aerosolized bird and bat droppings, in demolition workers. By the early 1970s, word about the refreshingly straightforward scientist from Cincinnati had gotten around. Bingham—likable and authentic, not averse to drinking in bars with rough union men—was asked by icons such as Selikoff and Mazzocchi to present at scientific meetings and speak at press conferences. She was an adviser to NIOSH, OSHA, the National Academy of Sciences, the Food and Drug Administration, and the EPA. In June 1976 she co-chaired a groundbreaking conference in Washington on health-related discrimination against women in the workplace. In her opening remarks she asked if the needs of female workers had ever been considered when new standards were being mulled, "or is the prototype of the American worker always the white American male?" Science showed that, in addition to any direct damage they inflicted, toxic substances could invade the womb, causing a woman to miscarry or bear children with birth defects or predisposition to illness. Diethylstilbestrol, or DES, a synthetic estrogen widely administered midcentury to prevent miscarriages, was a prime example. A breast carcinogen, it also was found to increase the risk of vaginal and cervical cancer in female offspring of women who'd taken it and the risk of testicular abnormalities in males. Lead was another example. It could short-circuit the worker's nervous system—perhaps even kill her in high doses—and cause severe mental impairment in her child. "Should the solution be not to hire [women], to fire them, to ignore them, or should it be to set an occupational standard that will protect them?" Bingham asked. In response, the OCAW's Sylvia Krekel neatly summarized the dilemma: "First of all, women are hampered that they are female, making it hard to break into the workplace initially. Once this hurdle is overcome and they get a job, the woman worker is the first to get laid off since she has the least seniority. With the addition of screening policies aimed at the fertile female, the position of women in the workplace becomes tenuous at best. The specter of triple discrimination becomes even more ominous if the woman becomes pregnant and the company decides she must transfer to another job for the sake of the fetus. She then loses the little seniority she may have acquired and usually must take a job at a lower wage scale." Panelists upbraided the government and debated the science. A few told stories worthy of Dickens or Sinclair. Jeanne Reilly, who worked in the lab at the Hooker Chemical plant in Niagara Falls, said the company had recently been fined $900 by OSHA for an explosion that killed four workers and propelled a cloud of chlorine into the surrounding neighborhood, sending ninety people to the hospital. Two weeks before the explosion, an apprentice welder had lost the top half of his head in a flywheel accident. Hooker, she said, made "many compounds whose chemical formulation the company will not tell to the employees." Workers were warned, for example, that Lindane, used in head-lice shampoos, was "bad for them. They weren't told that it was extremely poisonous, and that it was absorbed by the skin."

Bingham closed the meeting by saying she was despondent that there were few, if any, women in policymaking positions at the Labor

Department. And yet, when a member of the Carter transition team called her about the OSHA post six months later, "I laughed at them," she recalled. "I said, 'I couldn't do that. I've got children.' I was divorced. I said, 'I just can't do it and I'm not interested.'" She nonetheless agreed to meet with Labor Secretary Ray Marshall, an economist who had joined the administration from the University of Texas. "I said, 'Look, I've got a million dollars' worth of grants and contracts. I can go anyplace in the world I want to go; why would I want to do this?' He said, 'You sound just like me, and here I am.'" Bingham took the job and quickly got to work, issuing an emergency temporary standard for benzene that lowered the exposure limit over an eight-hour workday from 10 parts per million to 1. A NIOSH study of two Pliofilm plants in Ohio, showing a five- to tenfold increase in leukemia risk for benzene-exposed workers, had sealed the deal. "I said, 'God damn it, we're going to have it,'" Bingham told me long after she retired.

Bingham held her first press conference, on April 29, 1977, in tandem with Marshall to discuss the emergency standard. The standard was needed, Marshall explained, to keep more than 150,000 workers exposed to high levels of the solvent from getting leukemia. "The need to act is urgent," he said in his thick Texas accent. "I believe this signals a new day for an agency which in the past has been criticized for acting too slowly when lives were at stake." OSHA, he said, would move with "all due speed on serious health risks." Bingham reinforced this idea after stepping to the lectern, saying, "We plan to issue standards which deal with whole families of chemicals as well as single substances. . . . This is only the beginning." She comported herself well during the question-and-answer session with reporters. Was 1 part per million safe? one wanted to know. "That is as low as is feasible," Bingham replied. How long would the temporary standard stay in place? Until a permanent one took effect.

Bingham said she believed that when a substitute for benzene existed, it should be used. "We really do not know whether there is a safe level," she said, noting that in her lab in Cincinnati, the decision had been made to switch from benzene to toluene in the 1950s. She correctly anticipated resistance from industry. It came in the form of a court challenge by the American Petroleum Institute, which led to a temporary restraining order halting enforcement of the temporary standard. OSHA weighed its options and decided to go ahead with a permanent standard, which languished in the courts for nine years before taking effect.

News reports about Bingham were mostly positive. A profile in the *Cincinnati Post* noted that her "breezy informality charmed a Senate committee that asked her only about 10 easy questions before confirming her." Her aim, she told the *Post* reporter, was to transform OSHA from a "sometimes nit-picking bureaucracy into a real force for worker safety." Marshall reinforced this idea in an interview on the *Today Show*, saying, "We're going to stop fishing for minnows and start going after the whales." Among the whales was lead, one of the best-known, and most poorly controlled, industrial hazards. It was ubiquitous in Rome from 500 to 300 BC, found in everything from coins to irrigation and plumbing systems, used as a condiment, a wine additive, and even a method of birth control. It was present to a lesser degree in medieval Europe, gave people colic and other ailments, and was first banned by the town of Ulm, Germany, in the 1690s. Its effects on the unborn were well documented by the late nineteenth century. In a lecture delivered in London in 1911, Sir Thomas Oliver, a physician, said he had been working for nearly two decades to get pregnant women out of white lead factories, knowing that lead "destroys developing life by directly poisoning it, or it checks the growth of the fetus in the womb by cutting off its channels of nutrition." He cited a fellow researcher who found that children lucky enough to survive a year were of "an inferior physique" to their peers and were consigned to "mental as well as physical deterioration" as they grew. No part of this wretched history dissuaded the United States from becoming the world's biggest lead producer in the early twentieth century, infusing paints with it and adding it to gasoline to keep engines from knocking. It permeated battery plants and construction sites. The standard in effect when Bingham took office allowed workers to be exposed to as much as 200 micrograms of lead per cubic meter of air, enough to make them loopy, impotent, and bilious. Bingham cut the limit by three-quarters, to 50, and established a policy requiring workers with blood-lead levels of 60 micrograms per deciliter or over to be removed from their jobs and given full pay. This was a radical notion at the time.

The lead limit was one of six health standards, covering some of the worst substances in American workplaces, issued in 1978 alone. A *New York Times* article in February of that year offered an assessment of Bingham's performance as she approached a year on the job. It opened this way:

> The Labor Department's Occupational Safety and Health Adminis-tration has traditionally been one of the most criticized and battered agencies of the Federal Government, with virtually every act greeted with howls of anger from industry and ample helpings of scorn from labor. Lately, however, the howls have subsided to dyspeptic mutters, and the scorn has given way to expressions of praise and admiration.
>
> Credit for this turnaround is given chiefly to a rather motherly looking 49-year-old, Eula Bingham, who became Assistant Secretary of Labor for Occupational Safety and Health in the Carter Admin-istration last March. In a town that does not award such accolades lightly, Dr. Bingham is rapidly gaining a reputation as one of the toughest Federal regulators.

The piece quoted Bingham as saying, "We have to catch up with 200 years of tragedy in the workplace. There are things we have known about for more than 200 years, things such as lead poisoning, dangerous fire escapes, dangers from aromatics and metals that we haven't done anything about. But we can catch up. And when we do, the burden on this agency will become much lighter." Christie Waisanen, a lawyer with the US Chamber of Commerce, one of OSHA's fiercest adversar-ies, told the *Times*, "We like Eula Bingham; we like her common-sense approach." (In 1978, for example, the agency would do away with 928 fatuous provisions that had made it the butt of derision. Two years later, it was writing 37 percent more "serious," "repeat," and "willful" cita-tions—the kind that got bad companies' attention—than it was when Bingham took office.) But Waisanen said the Chamber disliked some of Bingham's initiatives, most prominently her proposed generic standard for carcinogens, an attempt to jettison the chemical-by-chemical approach that was excruciatingly slow and almost always led to industry lawsuits. The policy would have required employers to set "lowest feasible" exposure limits for potentially cancer-causing substances, which would be listed in two categories: those for which human or animal evidence of carcinogenicity was strong, and those for which the evidence was merely suggestive.

On July 11, 1978, a group of Steelworkers officials appeared at a public hearing in Washington to speak in support of the cancer policy. Among them was Mike Wright, a young industrial hygienist who would go on to become the union's director of health, safety, and environment. Wright allowed that the policy could lead to overregulation of carcinogens in some cases. "I think the people in this country generally would like to see those mistakes, if they are ever made, [be] made on the side of safety. We would rather see companies occasionally spend a little money unnecessarily than see workers suffer and die unnecessarily." Wright noted that "ever since I was in school, I have heard the words that America is a great country, and the reason why is we have a great economic system which provides the greatest standard of living in the world, and I think that is true. Well, given that kind of statement that we hear all the time, it puzzled me in these hearings to hear that in order to protect that great economic system we had to put up with 3,000, 20,000, 50,000 cancer deaths a year from occupational causes in order to protect this great standard of living. Now, it seems to me there is a real contradiction there. It seems to me that people who talk about risk-benefit analysis are saying cancer is a disease we cannot afford to live without. I think this country is great enough to live without occupational cancer."

Wright told the story of John Snow, a London physician credited with ending a cholera outbreak in that city's Soho neighborhood in 1854 by convincing authorities to disable a pump on Broad Street that drew sewage-tainted water from an especially foul stretch of the Thames. In what is considered the genesis of epidemiology, Snow conducted a study that showed people who drank from the well had a cholera rate fourteen times that of people who consumed cleaner water from upriver.

In a wry nod to the regulatory thicket that had formed in the 1970s, Wright noted that Snow didn't say, "Well, this is only one study," and insist on looking for exculpatory data for the bad well water. He did not have to wait for "the 19th-century British equivalent" of the National Academy of Sciences to bless his work, nor did he say superciliously that life was full of risks—"being run over by a horse-drawn shay . . . being mistaken for a pickpocket and being hanged"—and everyone obsessed with polluted drinking water should just get over it. "He did not try to put a number on the value of life and say, 'We have to figure out what regulation would be consistent with this particular value of life and how much it is going to cost and that kind of thing.' He did not do any of the things that I just outlined. What he did was, he broke the pump on Broad Street."

Wright suggested modern-day regulators could borrow a page from Snow and, in their own way, "break the pump." He did not mean "we should break industry. All I am saying is that we should control things down to the lowest feasible level." Wright addressed the empty threat of plant closure in the context of an arsenic standard for which the Steelworkers had lobbied, mostly on behalf of highly exposed smelter workers who were dying of lung cancer at excessive rates. "We ran into companies saying that if OSHA came in and did this, they would be forced for economic reasons to close down the plant," he said. "Well, OSHA has promulgated an arsenic standard. The plant where we heard most of those threats has not closed and in fact is working on a compliance program."

David Wilson, president of the Steelworkers local at Bethlehem Steel's Sparrows Point plant, was more strident when he testified that evening. Wilson had begun working at the plant in 1951 and said that the five thousand workers he represented were dying of cancer in alarming numbers. "The history of occupational health in this country has shown that industry first ignores and denies the existence of occupational disease, and when this is no longer possible, industry attempts to shift the blame and to delay indefinitely having to take measures to eliminate the problem." One vehicle for "endless delay," he said, was the "substance-by-substance" approach to chemical regulation. In 1962, as recording secretary of the local, Wilson had begun tabulating deaths at Sparrows Point and found that 43 percent of active members had died before they turned fifty-five. The average age of death for active and retired steelworkers combined was 63.5. By 1973, three years after he became president of the local, "the evidence was overwhelming that the people I represented were going to their graves before their time," Wilson said. Bethlehem Steel at first had "refused to cooperate with the union" in investigating the underlying causes of this loss of life, then was shamed into conducting its own mortality study after a Johns Hopkins researcher found a 56 percent excess of lung cancer deaths at Sparrows Point and a 100 percent excess of fatal esophageal cancers. Bethlehem's own study—which, like the Hopkins study, "suffered from methodological weaknesses" because of limitations in company-provided data—also found excess deaths, Wilson said. The findings of both studies, which almost certainly underestimated the true extent of slow-motion annihilation, were confirmed by NIOSH.

Wilson attacked the principle of "socially acceptable risk" then popular in more than a few corporate boardrooms. "Just as slavery was not socially acceptable, just as Hitler's extermination of the Jews was not socially acceptable, neither today is the genocide of the class of people called the American worker socially acceptable." Some industry leaders, worried about the bottom line, had complained at the hearing that OSHA was "overstepping its bounds" with the cancer policy, Wilson said, but "OSHA is precisely within its bounds. . . . No law on our books says that economic considerations excuse manslaughter."

The cancer policy, reviled by industry, never took effect. It was a lost opportunity, a sensible approach to an intractable problem that would not be proposed again. One wonders if it might have made a difference at a place like Goodyear, where the bladder-cancer cluster would emerge a decade later. Industry, however, wasn't the only barrier to progress. Marshall and Bingham sometimes clashed with Carter's chief economic adviser, Charles Schultze, who worried about the effects of overregulation. There were reports in the summer of 1977 that one of Schultze's ideas—replacing harsh punishment for employers who brooked unsafe workplaces with economic incentives to do better—was under consideration. Labor leaders were incensed; in a telegram to Marshall, OCAW president A. F. Grospiron said the union's experience had shown that "only stringent safety standards enforced on the plant level" did any good. "It is hard for us to conceive of any 'economic incentive' that the government could institute against these corporations, some of which are the richest and most powerful in the world, that could change the situation."

Things came to a head over the cotton-dust standard, designed to protect textile workers from byssinosis, better known as "brown lung." OSHA was under a court order to produce one, and Bingham—moved by an emotional meeting she'd had with wheezing members of the Carolina Brown Lung Association—complied, drawing the ire of textile executives by insisting that they install machinery to tamp down the dust rather than simply hand out respirators. She and Marshall were summoned to a meeting at the White House to defend the rule, which Schultze claimed could bankrupt the industry. Bingham had called Selikoff beforehand and asked what she should do. *Hold your ground and quit if necessary*, Selikoff advised. In the Oval Office, with Schultze looking on, Bingham appealed to the engineer in Carter, explaining that workers wouldn't be safe unless the dust was suppressed. The tactic worked. Carter suggested a compliance timeline even more aggressive than the one Bingham had proposed. Bingham was astonished. She knew Carter had taken campaign contributions from textile manufacturers when he ran for governor of Georgia and president. But she also knew he was principled. Bingham had less luck with Carter on beryllium, the metal that devoured workers' lungs and was used in nuclear weapons production. She and Marshall pushed for a standard but were outmaneuvered by the secretaries of energy and defense, who argued that the substance was vital for national security.

Another blow came after the revelation that the American Cyanamid Company had insisted that female workers of childbearing age at its chemical plant in Willow Island, West Virginia, be sterilized or face termination. This was the company's twisted way of protecting fetuses against lead released during certain processes. Betty Moler was among five women in the plant's pigments division coerced into having the procedure, called laparoscopic tubal sterilization, in which the fallopian tubes are sealed to prevent fertilization of the egg. "I remember everything about it, to the hour," Moler told the *Chicago Tribune.* "On January 30, 1978, they called us into the medical department. I even remember the time: 3:30 p.m." The plant's manager, doctor, and nurse briefed the women on what they euphemistically called "Band-Aid" or "belly-button" surgery. Moler tried to talk her way out of it, explaining that her husband already had had a vasectomy. The company people were unmoved, arguing that she still could get pregnant. Moler had the procedure, which did not go smoothly. Promised she'd spend one day in the hospital, she stayed six. Post-operative complications kept her out of work for six weeks; the company covered her medical costs and kept her on full salary. In retrospect, Moler told the *Tribune* in a quavering voice, "I didn't do the right thing. I know that now. I should have waited. I should have fought. That operation took a lot out of me." Because Willow Island was an OCAW plant, Wodka landed in the thick of things. He and two of the women appeared on a *Phil Donahue Show* segment that aired January 15, 1979. All the national newspapers published stories, and Wodka was bombarded with interview requests. After his *Donahue* appearance, he received letters of thanks—and condemnation. H. Christine Whiteman of Oklahoma City asked "why 'rights' and 'freedoms' so often appear to be the sole properties of the accusers. Has some group, agency or government denied these women the 'right' or 'freedom' to seek employment in some area not involved in high-risk production? . . . One cannot help appreciating the patience of business management and their reluctance to simply 'close the shop' forever."

Bingham saw the Cyanamid policy for what it was: illegal discrimination and an abdication of responsibility. OSHA cited the company for violating the general duty clause of the Occupational Safety and Health Act, a catchall provision inserted in the law because Congress knew it would be impossible for OSHA to issue a standard on every conceivable hazard. The company contested the citation administratively and in court, winning both times. Judge Robert Bork, then on the DC Circuit Court of Appeals and later a failed US Supreme Court nominee, and two colleagues held that the sterilization policy was not a hazard under the act. Any harm the women experienced came from "economic and social factors which operate primarily outside the workplace," Bork wrote in the circuit court's unanimous opinion. "Were we to decide otherwise, we would have to adopt a broad principle of unforeseeable scope" that could beget limitless liability for employers. Legal experts and labor leaders denounced this narrow interpretation of the statute, saying it would keep OSHA from addressing transplacental poisons and otherwise hobble the agency. In 1983, Cyanamid settled for $200,000 a lawsuit brought by the American Civil Liberties Union on behalf of the women who'd been sterilized and seven others who were transferred or demoted after refusing to undergo the procedure. For those who felt Cyanamid should have been punished harshly to send a message to other companies with exclusionary practices, the settlement was weak tea.

Bingham's time at OSHA wasn't just about cathartic policy reform, though there was a lot of that. She was the first leader of the agency to fully grasp the value of publicity, drawing on the cleverness of public relations adviser Frank Greer. One OSHA brochure decried the "massive yet silent slaughter" of at least one hundred thousand American workers a year from chemical exposures. Another featured on its cover the doleful, wrinkled mug of Louis Harrell, a North Carolinian who died of brown lung in 1978. (The latter so inflamed Bingham's successor that he ordered it pulled from circulation and demanded that warehouse stocks be destroyed.) Bingham commissioned and distributed edgy films such as *Can't Take No More* and *Worker to Worker*, narrated by oral historian Studs Terkel. Viewers at union halls and on college campuses saw grainy, archival footage of early-century explosions, fires, and mine cave-ins, and interviews with modern-day workers who spoke of enduring cancer, dust-caked lungs, and other ailments destined to kill them. At one point, audiences heard Johnny Paycheck's country anthem, "Take This Job and Shove It," which cannot have pleased the business establishment. The films, like the brochure featuring Harrell, would be banned after Carter was vanquished by Ronald Reagan. In fact, most of Bingham's ambitious agenda was scrapped after the election and replaced with stasis and backsliding. Bingham died in 2020, having remade, temporarily, an agency that had become "something of a laughingstock," the *New York Times* reported in her obituary.

CHAPTER 9

A BLUE-COLLAR SOCIAL CLUB

THE 1950S WERE NIAGARA FALLS' GLORY YEARS. The stores on Main Street were thriving, as were the factories on Buffalo Avenue. The population exceeded ninety thousand. There were six hotels, eight movie theaters, two hospitals, and two golf courses. It was during this time that Henry T. Schiro, a funny, sociable man who went by Hank, was hired at the Pathfinder chemical plant at a salary of $56 per week, or about $3,000 per year. Schiro had gone as far as the tenth grade at Trott Vocational High School and worked at a paper mill, a company that printed business forms, and in a warehouse before joining Pathfinder in December of 1952. Pathfinder, owned by Goodyear, was part of the parent company's inexorable push into organic chemicals and plastics, an industry it predicted would triple or quadruple in size by 1975. Contrary to public perception, Goodyear was not "a tire company doing some chemical business on the side," as a *Chemical Week* reporter put it in a 1960 memo to an editor. It and its counterparts had invested lavishly in research and development, turning out products that allowed "the average American today [to live] better than a king in the last century," a Goodyear chemist boasted in a 1962 speech. Schiro rode the R&D wave, starting as an operator in Niagara Falls and finishing as a maintenance mechanic fifteen years later. He worked mostly in Department 245, where the antioxidant Nailax was made, and became president of OCAW Local 8–277. He encountered ortho-toluidine as he repaired leaking pumps, though there were no cancer warnings posted and he would learn nothing about the chemical until he attended a medical conference years later. He wore a full face shield and rubber gloves when he was making repairs to keep from getting burned. Otherwise, he was unprotected against fumes and skin-penetrating liquids.

Schiro left Niagara Falls in 1967 to take a job as an OCAW international representative in New Jersey, responsible for the locals in that state. In 1974, the year before he was reassigned to western New York, he was diagnosed with bladder cancer, the second such case from Goodyear (though for many years his was thought to be the first). He was forty-five years old; the ginch had found him after twenty-two years in latency. "I went to the bathroom one evening and I passed a slight clot," Schiro explained in a deposition. His doctor ordered a cystoscopy, in which a thin tube fitted with a lens is inserted into the urethra through the opening of the penis. The procedure detected a tumor, which was removed and found to be cancerous. Back in Niagara Falls in 1975, Schiro had a follow-up cystoscopy that showed inflammation but no tumors. Come back in three months, he was told. It was the last time Schiro was hopeful. Eighteen subsequent penile incursions yielded discouraging news. "Every time [the doctor] went in there he would find a tumor, which he would have to excise," Schiro testified. Afterward, the pain that came with urination was almost unbearable. Sex with his wife, Dorothy, became infrequent, shattering his self-esteem. His doctor tried to kill the tumors with five rounds of chemotherapy. "He would go into the penis with a tube and stick approximately a quart of fluid in there . . . and request me to hold it for approximately two hours," Schiro recalled. It didn't work.

As Schiro received his ineffective and supremely painful treatments, twenty-one-year-old Harry Weist was getting started on the vinyl side of the Goodyear plant—Department 145. The two would never meet, but Harry, years later, would get a taste of what Schiro had gone through. Harry was bagging PVC resin, a job he secured in December 1977 through his mother, Jeanie, a switchboard operator with the company. Each day, Harry and a partner were expected to fill between eight hundred and a thousand fifty-pound bags with the resin, which came in different grades:

some were as smooth as flour and used in the production of surgical gloves; others were coarser and fabricated into car dashboards and vinyl flooring. Harry quickly learned the workflow in F Building: one man would bag for half an hour while the other took an unsanctioned break; most of the supervisors didn't care as long as the quotas were met. It was here that Harry met Robert Dutton. Alike in temperament, they became inseparable and collaborated to thwart boorish behavior by the few bad bosses. Their names wound up on a confidential list of employees marked for termination by an ill-tempered department head; Harry found the list, alerted the union, and that was the end of it.

Goodyear was not an oppressive place to work. The plant was a sort of blue-collar social club, where workers tormented one another with practical jokes, threw money into betting pools, bowled, and organized outings to Buffalo Sabres hockey games and Buffalo Bills football games. Opportunities for overtime pay were plentiful. One eight-hour shift at regular pay might be followed immediately by another at time-and-a-half, meaning one could get twenty hours of pay for sixteen hours of work. There was a pecking order, but young workers were welcomed into the fold, assuming they were reasonably competent and good-natured. A few days after he started, Harry was offered a pre-Christmas drink by a group of old-timers in the break room. He asked for and received a Seven and Seven, then nervously slid it away when one of the big bosses walked in. He needn't have worried; the boss accepted a drink as well. Harry heard from his elders how things used to be in Department 145, before vinyl chloride was stigmatized as a carcinogen. Reactor cleaners knew to exit the vessels when they got dizzy from the fumes, but there was no mention of anything more serious. The prevailing sentiment was that you shook it off and got back inside. At least two cleaners at Goodyear died of angiosarcoma, the rare liver tumor whose emergence at PVC plants triggered all the government activity in 1974, and at least one died of brain cancer. (A fourth worker, who'd produced vinyl chloride monomer, which the plant stopped making in 1967, also died of angiosarcoma.) Harry, who didn't enter reactors and benefited from a stricter plant-wide vinyl chloride limit, was spared.

For all the trouble it caused, vinyl production was a mainstay of the Goodyear plant for fifty years. Headquarters bragged about it incessantly. Publicity photos showed a bespectacled engineer in Niagara Falls admiring tiny cubes of PVC suspended around him like confetti, and a blond toddler ("Young Mr. America") standing in a playpen made of heavy-duty vinyl sheeting. In a press release announcing a $3 million plant expansion in 1967, Goodyear said it "expects the use of polyvinyl chloride industrial and consumer products, which exceeded two billion pounds annually in the United States for the first time last year, to increase about 10 percent a year as new uses are developed for it. Among the new uses are plastic bottles and plastic pipe for household plumbing and industrial transmission lines." After a setback in 1975—"economically unfeasible" workplace-exposure rules had forced a 50 percent cut in resin production—the company reported the following year that a new unit would get things back on track and add twenty jobs to the plant's 350-person payroll. A 1979 release quoted J. David Wolf, general manager of Goodyear's Chemical Division, as saying, "We feel the health issue is behind us now, and even with increased costs due to government safety regulations, PVC is still priced competitively with other plastics." Major markets for the resins included "flooring, auto parts, toys, foam padding, luggage and sporting goods." Inside Department 145, the "health issue" rarely was discussed. Premature deaths were treated as unfortunate coincidences.

After four years or so of bagging resin in F Building, Harry began working as a utility operator in Department 145. He often ventured into the C-2 building in Department 245 to use a steam hose or other equipment and was repulsed by the stench, which clung to and ruined his winter coat. His lips would tingle. Both were signs of exposure to ortho-toluidine, a chemical with which he wasn't familiar. In Department 145 he would charge the PVC reactors in the E-2 Building, filling them with water, catalyst, and liquid vinyl chloride and heating the mixture for about three hours until it looked like a giant vat of milk. Steam and gaseous vinyl chloride were removed by a machine called a degasser, and the batch was piped over to Building F, where it was de-watered in a centrifuge, put through a rotary dryer, bagged, put on a pallet, and sent to the warehouse for shipping. E-2 was plagued by vinyl chloride leaks during this time. Alarms went off at 5 parts per million, the legal short-term limit, at which point a plant-wide announcement would be made over the intercom and a lab technician would be dispatched with a back-mounted sensing device to sniff out the source. Not infrequently, the readings would cause the meter to spike, suggesting broken pumps or failed gaskets were spewing heavy concentrations of the carcinogen. One of the technicians who did the sampling in the summer of 1978 was nineteen-year-old Diane Kline, a seasonal employee whose father, Ray, worked in maintenance at the plant. Her forays into E-2 put her in regular contact with Harry, then known, somewhat unfairly, for his partying, quick temper, and general insouciance. The two became friends despite their dissimilar upbringings—Harry was undisciplined, from a troubled family; Diane, a diligent student, close to her parents, and had belonged to her high school Bible Quiz team.

That summer, Niagara Falls made news for the wrong reasons. The city's reputation had already sagged after a series of rockslides all but destroyed the Schoellkopf Power Station in 1956 and companies such as Union Carbide and Carborundum began shedding workers in the early 1960s. "Much of the downtown area of the city looks old and tired," the *New York Times* reported in 1963, quoting several citizens anonymously so they could freely express their ennui. "We've been drifting downhill for a long time," a local banker observed. Four years later, the *Globe and Mail* of Toronto used the Niagara Falls Urban Renewal Agency's own words in characterizing the city as "shabby, forlorn and uninspiring." In the early 1970s Mayor E. Dent Lackey, a former Methodist minister known for his upbeat manner, exclaimed to an Associated Press reporter, "We're tearing it all down and building it all over again, but, oh God, what a fight it's been!" There was talk of constructing the world's first rotating hotel (didn't happen) near the American Falls and hosting the 1976 Democratic National Convention (again, no). Eighty-two acres of downtown blight were leveled and a convention center, co-designed by Philip Johnson, took shape, but the hoped-for knock-on effect never materialized. There were rumors of graft and backroom deals.

In 1978, the *Niagara Gazette* published a series of articles on hazardous waste buried beneath the Love Canal neighborhood, a few blocks
from the Niagara River—a "full-fledged environmental crisis," reporter
Mike Brown called it. Hooker Chemical, acquired by Occidental in
1968, had been dumping exotic slurries and dregs into the unfinished
canal since 1942, believing its clay soil would keep the poisons from
migrating. Houses and a school were built atop the landfill in the 1950s.
Storm sewers were cut into the clay, and road construction in the late
1960s further compromised the dump's structural integrity. By the late
1970s homeowners were complaining that chemicals had seeped into
their basements, and heavy rains had destroyed what few barriers to
unbridled contamination remained. "Corroding waste-disposal drums
could be seen breaking up through the grounds of backyards," Eckardt
Beck, then the EPA's regional administrator in New York, wrote after
touring the area. "Trees and gardens were turning black and dying. One
entire swimming pool had been popped up from its foundation, afloat
now on a small sea of chemicals."

Lois Gibbs, who grew up in insular Grand Island, New York, and
her husband, Harry, moved to Love Canal in 1974 with their young
son, Michael. It was a subdivision of ranch houses in which most of
the men worked at the plants—Harry Gibbs at Goodyear—and most
of the women stayed home and raised their children. The Gibbses had
a daughter, Melissa, and settled into the rhythms of a neighborhood
accustomed to shift work: 3 p.m. to 11 p.m., 11 p.m. to 7 a.m., 7 a.m.
to 3 p.m. People watched one another's kids, carpooled, and otherwise
cooperated. The 99th Street School was at the center of it all. Michael's
health deteriorated as he finished kindergarten there in the spring of
1978; he developed asthma, seizures, and problems with his liver, urinary tract, and immune system. His mother was flustered but didn't
associate any of it with chemicals from the crumbling dump site. Then
she began reading Brown's *Gazette* articles and made the connection.
Gibbs overcame her natural shyness and assumed the presidency of the
Love Canal Homeowners Association, which demanded the closure of
the school and the relocation of residents. She went door to door, collecting signatures on a petition and, on occasion, being shooed away.

On August 2, 1978, New York State health commissioner Robert P. Whalen declared a state of emergency, ordering the school shuttered and recommending that pregnant women and children under two evacuate the neighborhood. A federal declaration by Jimmy Carter five days later provided funds to relocate 239 families that lived in the first two rows of homes encircling the landfill; those outside that ring, including the Gibbses, were told they were not at risk and got no such relief, despite evidence of excessive rates of birth defects—extra fingers and toes, double rows of teeth—miscarriages, and stillbirths. That decision was reversed by Carter in May 1980, and another 710 families were made eligible for subsidized relocation. In December of that year, shortly before he left office, Carter signed into law the Comprehensive Environmental Response, Compensation, and Liability Act, a toxic-waste cleanup initiative better known as the Superfund law, which became Love Canal's best-known legacy. Lois Gibbs, who started an advocacy group called the Center for Health, Environment and Justice in 1981, maintains that there was an equally important, lesser-discussed takeaway: the revelation that exposure to low levels of chemicals, over time, can do severe damage to the human body. Until Love Canal, Gibbs told me, most bought into the industry trope that "the solution to pollution is dilution." At the peak of the crisis, her husband told the *Gazette* he felt "safer at Goodyear than in my own home. At least I was sure at Goodyear of the chemicals I was exposed to."

In the summer of 1979, as the Love Canal imbroglio was unfolding, Diane Kline returned to Goodyear as a full-time lab employee, abandoning her studies at Erie County Community College. She and Harry Weist began dating. "He had a reputation as being kind of a hell-raiser, but he wasn't a bad guy," Diane recalled. "He would never hurt anybody, but he would stand up to bullies. He wouldn't take crap from anybody." Ray Kline objected to the relationship at first, but Diane pushed back and Ray softened. Diane and Harry were married at the Klines' family church, the Christian and Missionary Alliance in Niagara Falls, on February 2, 1980. Diane took disability leave from Goodyear in August of that year after she learned she was pregnant. She remembered what had happened to her brother and sister when she was young and wanted no part of the chemicals in the lab, despite pressure from the company to return. Harry was fearful as well and insisted that Diane weather the pregnancy—which sent her to bed with toxemia for six months—at home. Diane gave birth to a healthy girl, Holly, on May 26, 1981, and never went back to Goodyear. Harry, who'd lived near Love Canal as a preschooler, was relieved. He and Diane planned to have more than one child, and he was uneasy with the thought of her working again around chemicals. He worked one, sometimes two, double shifts each week to make up for the lost income.

CHAPTER 10

DUPONT AND DOMINIC

THE FISHY-SMELLING CHEMICAL known as ortho-toluidine, which caused so much misery at Goodyear, is made by reacting nitric acid with toluene, an organic solvent first isolated through the distillation of pine oil by Polish chemist Filip Walter in 1837. Light yellow at room temperature, it darkens as it's exposed to air and light. Apart from its function as a rubber antioxidant, it is used to make dyes, pigments, herbicides, and Prilocaine, a cream used to numb the skin during dental and surgical procedures and blood draws. Ortho-toluidine no longer is manufactured in the United States, but for much of the twentieth century was made by—among others—the E.I. du Pont de Nemours and Company, Goodyear's primary supplier in Niagara Falls from 1957 to 1995. Founded in Wilmington, Delaware, in 1802, DuPont had made its initial mark selling gunpowder, which proved immensely profitable, especially during the War of 1812. It went on to make industrial blasting powders during the second half of the nineteenth century. It opened a dynamite plant on the Delaware River in Gibbstown, New Jersey, some twenty miles southwest of Philadelphia, in 1880, and by 1902 had added an explosives research laboratory. Here were concocted dynamites that could be used safely in the extreme cold of northern quarries or amid the ignitable dust of coal mines. In 1891, DuPont bought a large tract of land downriver in Carneys Point, New Jersey, where it made highly flammable nitrocellulose, also known as guncotton, for naval mines and torpedoes before moving on to more pedestrian chemicals.

DuPont began producing ortho-toluidine, an aromatic amine, at its sprawling Dye Works (to be renamed Chambers Works) in Carneys Point in 1919. More than two decades earlier, in 1895, German surgeon Ludwig Rehn had reported finding bladder cancer in three workers at a dye factory in Frankfurt. Rehn noted that gases associated with the production of fuchsine—a deep red dye better known as magenta—caused "disruptions in the urinary tract"; that long-term work with fuchsine could induce bladder tumors; and that the oily liquid aniline—in structural terms, among the simplest of the aromatic amines—was the likely agent. Rehn, one of the early users of cystoscopy to search for tumors, had documented thirty-eight cases of bladder cancer in seven factories by 1906. German law eventually would force dye operations to improve ventilation, provide workers with protective clothing, and mandate post-shift hot baths. In a 1921 monograph, the International Labour Office in Geneva summarized the findings of Rehn and others in Europe, deeming it "absolutely necessary that in factories in which workers are exposed to the dangerous action of aromatic bases, the most rigorous application of hygienic precautions should be required." DuPont, by all evidence, did little or nothing in response. The Europeans—mainly the Germans—had dominated synthetic dye-making since the late nineteenth century, and after World War I DuPont and other American companies moved to fill the void created by Germany's defeat. "A shortage of dyestuffs threatened to throw millions of employees in the textile industries out of employment," DuPont vice president W. S. Carpenter Jr. said in a 1935 speech to foremen at the Dye Works. The United States also was "helplessly dependent" on Europe for medicines, anesthetics, organic chemicals for industry and agriculture, and "many of the indispensable materials of modern warfare," Carpenter said.

The Dye Works registered its first major health scandal in 1925 around the production of tetraethyl lead, an anti-knock gasoline additive. A *New York Times* article by Silas Dent on June 22 of that year reported that eight workers had "died in delirium" after being poisoned by the compound during the previous eighteen months and another three hundred had been "stricken, but not fatally." Employees had come to know the tetraethyl lead plant as "The House of Butterflies," Dent wrote, in acknowledgment of one of the oddest and most disturbing symptoms of poisoning: "a hallucination of winged insects." The first to die, thirty-seven-year-old Frank (Happy) Durr, was "ill only a short while" and expired in a straitjacket. DuPont admitted there had been poisonings during the plant's experimental phase but said it had shut down the operation in the fall of 1924 and adjusted the ventilation system to allow an exchange of fresh air every forty seconds. There were only "slight difficulties" afterward, President Irénée du Pont insisted, but Dent had found three more deaths. A DuPont spokesman dismissed as "absurd" suggestions that the deaths had been covered up.

Bladder cancer made its appearance in Carneys Point in 1929. DuPont physicians began documenting cases among dye workers and would find at least eighty-three by 1936. The suspect chemicals were benzidine and beta-naphthylamine (BNA). Production of the latter began in 1919; an internal company document described the process: "It was cast in open pans, broken with a pick, and transferred by hand into barrels, ground in an open mill, and shoveled by hand into operating equipment. There was no ventilation provided. Gross exposures occurred." In 1934, DuPont's medical director, G. H. Gehrmann, emphasized the importance of giving each highly exposed employee at the Dye Works an annual cystoscopy, the cringe-inducing procedure with which so many Goodyear workers would become familiar. Medical examinations, Gehrmann wrote, should "continue all through the entire period of employment, and in the case of men exposed to bladder tumor-forming chemicals, continue until death removes the final possibility of tumor development." Gehrmann, evidently shaken by the outbreak, recommended that DuPont open a research lab to test its growing line of chemicals on rats, guinea pigs, and rabbits, whose reactions were often predictive of human effects. The doctor's suggestion was taken, and the Haskell Laboratory for Industrial Toxicology, among the first of its kind in the United States, was dedicated in a Wilmington suburb on January 22, 1935. A DuPont-prepared retrospective touted the accomplishment twenty years later:

> Like people, chemicals are of many types and dispositions. Most are uncomplaining, law-abiding citizens who present no special problems.

Some are unstable and unpredictable. Some are out-and-out neurotics, requiring a sharp and observant discipline. Others, harmless by themselves, may be influenced by bad company. Still others are troublemakers in solitary, but docile and helpful in tandem. Some are just plain bad actors. DuPont and the chemical industry, more than aware of such tendencies, take extreme measures to screen and classify the offenders and keep them under proper restraint. Nor does their responsibility cease at the point of arrest and indictment. Unlike the law enforcement officer who merely seizes contraband, the industrial scientist's role may be likened more closely to rehabilitation. He prescribes the conditions and areas under which hazards are minimized and perils avoided; at the same time he permits the broadest range of usefulness.

Notable among the early Haskell scientists was pathologist Wilhelm C. Hueper, who was born on November 4, 1894, in the pastoral Mecklenburg-West Pomerania region of northeastern Germany, was drafted when World War I broke out twenty years later, served as a medic in a field hospital in France, and emigrated to America in 1923. Practicing at Mercy Hospital in Chicago, where he worked for frugal Irish nuns, Hueper began writing about environmental cancer, taking the contrarian view in a 1926 paper that cigarette smoking—then rising in popularity—couldn't be blamed for a lung cancer spike in central Europe. More likely, he theorized, the surge was due to automobile exhaust, coal tar and asphalt dust from roads, and smoke from furnaces and chimneys. Hueper was recruited to join the University of Pennsylvania's Cancer Research Laboratory—which received financial support from Irénée du Pont—and moved to Philadelphia with his wife, Martha, and two-year-old son, Klaus, in 1930. The lab was overseen by Dr. Ellice McDonald, an associate professor of gynecology at the university and the du Pont family's personal physician. Hueper pursued a new interest—"the biochemical problems of cancer growth," as he put it in an unpublished autobiography. He tried and failed to induce cancer in mice by dosing them with organic arsenic. He succeeded in inducing leukemia by exposing the rodents to ionizing radiation, which comes from sources both natural (radon gas, a decay product of radium) and human-made (X-rays).

One day early in his tenure, Hueper accompanied McDonald on a trip to the DuPont Dye Works, across the river in New Jersey. "During this visit," Hueper wrote decades later, "I was told by the DuPont chemists that in some of the aniline dyes, certain aromatic amines are used which I knew, from my previous abstract work of the German literature, were suspected of having caused cancer of the bladder in a high percentage of the exposed workers in Germany and Switzerland." After returning to Philadelphia, Hueper drafted a memo to Irénée du Pont, warning that the same hazard existed for American dye workers, who, like their European counterparts, were exposed to benzidine and BNA. He received no response from the company's president but heard from McDonald several months later that "there were no such complications among the American workers." Six months after that, Hueper wrote, medical director Gehrmann "appeared at our laboratory one late afternoon and declared he was distinctly alarmed that they had found 23 cases of bladder cancer" among DuPont workers. Hueper was not surprised. The company had begun producing aromatic amines some fifteen years earlier and the latency period matched what had been seen in Europe. "A new industrial cancer of occupational origin had arrived on schedule in the United States," Hueper wrote. Gehrmann's account of this period is similar, differing slightly in the details. DuPont, he reported in published papers, had seen three cases of bladder cancer by 1931 and twenty-five by the time he presented at a 1934 symposium on the disease. All twenty-five workers had been exposed to either BNA or benzidine.

Hueper came to know the du Pont family and at one point suggested to patriarch Irénée that the company establish a "central toxicological institute" for the study of chemicals as a means of protecting both workers and consumers. DuPont pondered the idea but passed, telling Hueper the expense associated with such a venture couldn't be justified in the depths of the Great Depression. Meanwhile, tension mounted between Hueper and his boss, McDonald. Hueper "took exception," he later wrote, "to [McDonald's] releasing a communication to the press before a meeting of the American Chemical Society, claiming that a new and important discovery as to the causation of cancers had been made in his organization and then continued to describe some details of the alleged experiments in support of his claim. The fact was that no such discovery had been made at our Laboratory and not even experiments on such a project had been started, because it was still purely speculative. I told him that through such irresponsible propaganda, he was ruining not his own reputation, but also that of all his associates." Hueper's truth-telling proved costly; he was fired by the university in 1934, searched without success for other work in Philadelphia, and returned with his family to Germany to visit his mother and brother and, he hoped, find employment. Several of his former university professors had discouraging news for him. For all their cruelty to humans, the Nazis had passed strict animal-protection laws. Hermann Göring, whose influence was exceeded only by Adolf Hitler's, had decreed an end to what he called the "unbearable torture and suffering in animal experiments." This spelled doom for a lab researcher, and the Hueper family sailed back across the Atlantic.

After a brief stint at a 250-bed hospital in Uniontown, Pennsylvania, a coal-mining town near Pittsburgh, Hueper landed a job as a pathologist at DuPont's Haskell lab, still under construction. Lab director W. F. von Oettingen "mentioned that they were especially interested in me, since I had discovered the bladder cancer hazard among dye workers and wanted me to work experimentally on this problem," Hueper recounted in his autobiography. The excitement of his new position wore off as he realized the experiments he and his colleagues would conduct were designed to wrap up before cancer or other chronic effects could appear. Not unexpectedly, most of these studies found no ties between chemical exposures and disease; an outlier, which lasted three years and involved the feeding of BNA to dogs, produced bladder cancer. Hueper came to conclude that commercial interests could not be trusted to protect the health of their employees or their customers. Two episodes stood out to him. The first occurred several months into the dog study, when he asked to see the BNA production area at the Dye Works. A manager complied, and Hueper was struck by the immaculateness of the operation, which had been idled for his visit. He complimented a foreman, who volunteered, "Doctor, you should have seen it last night. We worked all night to clean it up for you." This, of course, defeated the tour's purpose. Hueper then asked to see the benzidine operation, which was in a separate building. "With one look at the place," he wrote, "it became immediately obvious how the workers became exposed. There was white, powdery benzidine on the road, the loading platform, the windowsills, on the floor, etc." Hueper complained about the sleight of hand in a memo to Irénée du Pont. "There was no answer," Hueper wrote, "but I was never allowed again to visit the two operations."

In the second case, Hueper caught wind of a planned DuPont press release, written for a local newspaper, in which Gehrmann and von Oettingen would be credited with identifying the agent responsible for the company's bladder-cancer outbreak. "Since neither of the two had a blessed connection with this particular research," Hueper wrote, "I looked upon the attempted misappropriation of credit as an objectionable use of power bordering on scientific theft." He met with the newspaper's editor and convinced him not to publish, an act of defiance that wrecked Hueper's future with DuPont. Gehrmann fired him, ostensibly for economic reasons, in November 1937 and later warned him not to publish anything related to his work at Haskell without the company's consent. By Hueper's telling, this preemptive censorship continued for the rest of his career. It followed him to William R. Warner & Company, a New York pharmaceutical company he joined in 1938: a terse letter from Gehrmann warned Hueper not to present the findings of his dog experiment at the upcoming International Union Against Cancer meeting. Hueper, lacking the resources to fight DuPont in court, relented. It also followed him to the National Cancer Institute (NCI), where Hueper took charge of the new Environmental Cancer Section in 1948. Hueper had published his seminal textbook, the 896-page *Occupational Tumors and Allied Diseases*, six years earlier with no response from DuPont. "I had the faint hope that by now they had got off my trail and would leave me alone," he wrote. It was not to be. Hueper was informed by letter a few months into the NCI job that he was under investigation for disloyalty to the United States. A friend had been interrogated by an

FBI agent, who said the allegation had come from the DuPont medical department, "which had accused me of being a Nazi." After presenting a vigorous defense with the help of a lawyer, Hueper was cleared, only to learn a few months later that Gehrmann had written a letter to the NCI's director claiming Hueper had shown "communistic tendencies." Gehrmann also insisted the main culprit in the bladder-cancer epidemic at the Dye Works—now Chambers Works—was BNA, not benzidine. This conveniently aligned with DuPont's economic interests; it had begun to scale back BNA production and would stop making it by 1955, though would continue to buy the chemical from others until 1962. DuPont didn't stop making benzidine until 1967 and kept using it until 1972. Hueper was vindicated in his belief that both chemicals were carcinogens. In 1948, he wrote, Salem County, New Jersey, home to Chambers Works, had by far the state's highest bladder-cancer death rate. This was still true in 1975, "strongly suggest[ing] that the epidemic has continued without significant reduction." By 1991, according to a DuPont memo, the plant had recorded 489 cases of the disease, 453 of which were deemed occupational in nature. These were attributed to early- and midcentury exposures to BNA and benzidine. Ortho-toluidine, tightly controlled at Chambers Works for nearly four decades, wasn't tested for carcinogenicity, even though DuPont had the capacity to do so.

In fact, ortho-toluidine had been implicated in animal and epidemiological studies generations earlier. Hueper mentioned the chemical in a 1934 review paper; six years later researchers in Japan reported that ortho-toluidine caused tumors in the bladders of rabbits that had been injected with it and rats whose skin had been painted with it. Two follow-up studies were confirmatory. In 1952, a company doctor in Great Britain reported a case of bladder cancer in a worker involved in the making of ortho-toluidine; the same year another British physician wrote in a medical journal, "It appears likely that ortho-toluidine may be capable of producing bladder tumours in man and in the dog."

In a deposition more than three decades later, the Haskell lab's retired director, John Zapp, admitted knowing by 1955 that ortho-toluidine had caused tumors in rodents. "Look, I don't care if a chemical gives cancer to rats if it doesn't bother the humans," he testified. "And I think that the rat is a poor indicator for bladder tumors." By 1955 ortho-toluidine already was classified as a "no-contact chemical" at Chambers Works based on its acute effects: burns, cyanosis (bluing of the skin), eye irritation, dizziness, shortness of breath. Any worker likely to encounter it was sealed in what DuPont called a "Chem-Proof Air Suit." DuPont could have done its own experiments on ortho-toluidine at Haskell in the mid-1950s had it been so inclined. It could have done so again in 1974, after it learned of a rat study by the National Cancer Institute that found, in DuPont's words, "significant levels of tumor formation with some animals at some feeding levels."

Cancer bioassays of rodents are reliable predictors of human disease; DuPont's scientists surely knew this. But the company didn't warn Goodyear and other customers that the substance might be harmful to workers until 1977. In a letter on January 12 of that year, it called ortho-toluidine a "suspected carcinogen" that should be handled with caution but added, reassuringly, that despite a half-century of production at Chambers Works, "we have seen no evidence that it ever caused cancer in any of our employees." This might have been technically accurate but was disingenuous: DuPont hadn't *looked* for cancer among its employees. The company also failed to mention its knowledge of a 1970 study from the Soviet Union that had found eight bladder tumors among eighty-one workers who made ortho-toluidine. DuPont had obtained an English translation of the paper in 1975. The DuPont letter in 1977 caused no consternation at Goodyear, which would continue to run its Niagara Falls plant in a fashion that, a NIOSH analysis later found, grossly exposed Department 245 production workers to ortho-toluidine for another seventeen years.

The manner in which DuPont and its customers, including Goodyear and Monsanto, handled ortho-toluidine could hardly have been more different. Workers who loaded the chemical into fifty-five-gallon drums, railroad tank cars, and tanker trucks at Chambers Works wore impermeable rubber suits and breathed supplied air. Drum-loaders had glove boxes as an added layer of protection. DuPont's customers were on their own. Monsanto's medical director, Emmet Kelly, had written to Zapp in 1955, noting that some employees exposed to ortho-toluidine were seeing blood in their urine and asking for guidance. Zapp handed off the request to another DuPont doctor, E. E. Evans, who assured Kelly that while DuPont regarded ortho-toluidine "with suspicion," it was considered "much less severe" than BNA. For reasons of cost and lab capacity, DuPont centered its animal research on new products. Ortho-toluidine, in the company's view, had not shown itself to be worthy of in-house investigation.

In his autobiography, Hueper condemned American industry for its "policy of silence" toward occupational and environmental carcinogens. It was, he wrote acidly, more concerned with beating back regulations and maintaining customers than protecting employees. It ignored experimental evidence. It made the same engineering mistakes, over and over. It sought to dispatch the messenger rather than the hazard. "Active and aggressive work in this field is like participating in a strange type of war, which, like any other type of war, had its cruelties and casualties," the old pathologist wrote some two years before he died. "Looking back over my life, it seems that at various occasions I was singled out for forceful removal from the scene by various types of assassination. The only type which was never attempted was physical murder."

Not quite seven months after Hueper's death in December 1978, one of his young admirers, environmental consultant Barry Castleman, felt compelled to respond to an op-ed in the *Washington Post*, in which Carl B. Kaufmann of the DuPont public affairs department congratulated the company for its transparency and integrity. Kaufmann gave as an example the Chambers Works bladder-cancer episode. "We made full disclosure in the medical journals, cleaned up the process, and took care of employees to the best of medical science's ability," he wrote. In his own *Post* op-ed two weeks later, Castleman challenged this self-serving narrative, pointing out that it was Hueper, not the higher-ups at Haskell, who had sounded the alarm not only about BNA but also about benzidine, which DuPont kept making long after its carcinogenic properties had been identified. Castleman told the story of a visit to Chambers Works by two British researchers in 1949. During a long drive after a plant tour, one of the visitors, a medical officer with Imperial Chemical Industries, asked DuPont's Gehrmann how he could be "so certain that benzidine is not causing any of the trouble." Gehrmann reportedly replied: "We here know very well that benzidine is causing bladder cancer, but it is company policy to incriminate only the one substance"—BNA. In the back seat of the car—seemingly asleep but in fact listening with his eyes closed—was Dr. Robert Case. A DuPont representative tracked down Case after Castleman's piece was published and asked if the retired doctor could confirm or deny the story. Case confirmed it.

In his op-ed, Castleman took offense at Kaufmann's claim that "most people in business have pretty much the same base of values as most of their critics." What, Castleman wondered, would Hueper have thought of such a statement? It was not hard to imagine. "He called them chiselers, the callous businessmen who saved a few thousand dollars on industrial hygiene engineering," Castleman wrote. "He railed at them for suppressing the deadly truth from their workers, with their 'flexible' front-men in medicine, law and public relations. Bill Hueper learned about business ethics and occupational cancer from the people who wrote the book. 'The only thing they understand is jail and bad publicity.'"

CHAPTER 11

CANCER ERUPTS
AT GOODYEAR

I N THE SUMMER OF 1978, workers at Goodyear had yet to see Hank Schiro's bladder-cancer diagnosis four years earlier as any sort of omen. The main worry at the plant was a slew of complaints about chest pain, consistent with angina, in Department 245. That summer, a young physician named Christine Oliver took an internship with the Oil, Chemical and Atomic Workers union in Washington. Oliver, whose father and paternal grandfather were doctors and whose mother was a nurse, grew up in Raleigh and Smithfield, North Carolina. She graduated from the University of North Carolina School of Medicine in 1970, spent a year as an internal medicine resident at North Carolina Memorial Hospital in Chapel Hill, and then took a year off to counsel young women as director of family planning for the Wake County Health Department at the peak of the abortion wars. She continued her residency at Montefiore Hospital and Morrisania City Hospital in New York City, finished in 1974, and practiced at clinics in the South Bronx and nearly-as-distressed Chelsea, a city near Boston. An ardent believer in prevention, she came to realize her financially stressed patients were unlikely to respond to sanctimonious lectures on smoking, exercise, or diet but could benefit from safer workplaces. After hearing Tony Mazzocchi speak animatedly at a conference in Washington in April 1977, she was inspired to begin a second residency, in occupational medicine, at the Harvard School of

Public Health. Midway through it, she took the OCAW internship as part of her training.

At OCAW plants during this period, movement on matters of health and safety began with the rank and file. If a local chapter of the union had strong leadership, as was the case at Local 8–277 in Niagara Falls, members were encouraged to report problems. If those problems couldn't be resolved by the local, Tony Mazzocchi and Steve Wodka would get involved. Wodka might be sent to do a plant tour, interview workers, and compile a list of demands for the company. If the company didn't cooperate, Wodka would file a complaint with OSHA on behalf of the local, follow the investigation, and represent the OCAW before the Occupational Safety and Health Review Commission if the employer contested any citations. Mazzocchi was true to the process. He once believed, he told a Canadian audience in 1978, that occupational disease was "an aberration," the product of "a dirty plant or malicious people." He now understood that "the drive to increase productivity in Western society is the key component. With that must go an increased disease rate and increased cancer. The only growth product in this part of the world—the US and Canada—is cancer. It's growing at 5 percent a year. The GNP should be so lucky."

It was Oliver's job that summer to perform on-site evaluations of the union's refineries and chemical plants and identify the poisons that were felling its members. She stayed on part-time through the fall and winter and made her first visit to Goodyear in Niagara Falls on March 2, 1979. Oliver met with members of the local and did a walk-through of Department 245, where the antioxidant Nailax and an accelerant called Kagarax A, which helped speed the rubber-curing process, were made by ninety-six workers over four shifts. In a report to Mazzocchi, she described areas in which workers were being assaulted by chemicals known to affect the heart, such as carbon disulfide. She noted strong fumes in the tank farm, from which raw materials were pumped to the Nailax and Kagarax A reactors in Building 32, and near carbon disulfide storage tanks, which were allowed to overflow into buckets. The two Nailax reactors would sometimes soak the workers with their contents; their reeking, saturated filters—known by the name of their manufacturer, Sparkler—were cleaned manually without the benefit of a local exhaust system. Oliver saw potential exposure to three cardiotoxins: carbon disulfide, nitrobenzene, and aniline, an ingredient in Kagarax A. After finishing her Harvard residency in May 1979, Oliver went to work for the OCAW full time. That October, an independent industrial hygienist of her choosing, Robert Gempel, surveyed the Niagara Falls plant, looking for chemicals that could induce heart disease. Gempel found airborne exposures to ortho-toluidine, carbon disulfide, and four other substances of interest; all concentrations were below federal limits. He cautioned, however, that his sampling was performed within a small window—three days—and that exposures may have been heavier in the past. He raised the possibility of absorption through the skin, especially near the Nailax reactors, where there were "no safeguards in place to prevent accidental overflow of raw materials." In fact, during the midnight shift on Gempel's last day of sampling, several gallons of ortho-toluidine had spilled onto the floor. And he noted that he had conducted personal air monitoring only of production workers, not maintenance workers, "who potentially could be subjected to high exposures." Oliver, meanwhile, began her own study. Urine samples she collected from exposed workers in Department 245 showed elevated levels of "urinary aromatic amines," a combination of ortho-toluidine, aniline, and nitrobenzene. This suggested skin absorption. Samples from controls, in Department 145 (vinyl), showed far lower levels. The incidence of chest pain also was higher in the exposed workers. Oliver wasn't looking for bladder cancer, though she knew ortho-toluidine had caused the disease in animals. "These cases were lurking in the background," she recalled, "and somebody put two and two together."

By early 1981, four current or former Department 245 workers—including Hank Schiro, now an OCAW international rep—had been diagnosed with bladder tumors. It was the first clear sign that an outbreak was under way. Coached by Oliver, Rod Halford, then president of OCAW Local 8–277, sent an urgent letter in March to plant manager J. A. Pearson, seeking industrial-hygiene monitoring results and data on cancer mortality and incidence. Pearson replied to Halford's successor as president of the local, Jim Ward, that the union's concerns had been relayed to corporate in Akron but that air levels of ortho-toluidine and another suspect chemical, aniline, were "well below" what the law allowed. Mortality studies of plant workers in 1973 and 1979 had not "addressed cancer of the urinary tract specifically," Pearson explained, and incidence data was spotty because of medical-privacy issues. In a memo to Ward three months later, Oliver identified ortho-toluidine as "the most likely cause" of the Department 245 outbreak, noting it was a "suspect bladder carcinogen . . . known to cause blood in the urine. It is important that we pursue this, not only for the individuals involved but also to clarify the role of ortho-toluidine as a human bladder carcinogen."

In a later deposition, Halford said he interpreted Goodyear's response to his letter to mean there was "nothing to worry about." He found this acceptable at the time, though the union had fought with Goodyear over the pace of ventilation improvements in Department 245. The fumes that came off the Sparkler filters were notorious for their vileness. The residue on the filters after a four- to six-hour reaction was "molten black," Halford explained, and could be up to six inches thick. It took two men to dump the filters' contents into a drum. "And all the while there is a tremendous amount of fumes, and this stuff was hot—I mean, real hot. . . . I've seen it catch on fire. Cleaning of the filter was . . . a terrible job, and I got an awful lot of fumes and smoke while doing that." The buildup of residue from the reactors would dim lights and create a tripping hazard on the floors in Building 32. In the early days, pressure inside the vessels caused blowouts, sending clouds into the atmosphere and leaving a film on workers' cars. Goodyear was "always adding fans" in the vicinity of the reactors, Halford recalled, and eventually installed enclosed control rooms for operators like himself. But the company's warnings over the years focused on the explosive and caustic properties of chemicals such as sulfuric acid. Hence the emphasis on gloves and goggles. This equipment offered no real protection against ortho-toluidine—Dominic—which clung to workers' bodies even after they showered and created silhouettes on their bedsheets. White underclothing had to be thrown away. White shirt collars darkened. Gloved hands were stained brown. The lunchroom walls turned black and had to be cleaned with ice scrapers.

During her investigation, Oliver had several meetings in Niagara Falls with Goodyear's corporate medical director, Clifford Johnson, who had assumed the post in 1965 and, along with the company's industrial hygiene staff, was responsible for employees' well-being. In depositions prior to and after his retirement in 1989, the doctor was pressed to explain what he had known about ortho-toluidine before the four bladder-cancer cases came to light in 1981, and how he had responded to the news. He came across as evasive and forgetful. He had served, for example, on an OSHA advisory committee in 1973 that had nominated fourteen carcinogens—including BNA and other aromatic amines that attacked the bladder—for regulation. He'd attended Manufacturing Chemists' Association meetings before and after the B.F. Goodrich revelations on vinyl chloride in 1974. These experiences apparently hadn't aroused his curiosity or concern when the four Niagara Falls cases surfaced.

Johnson acknowledged learning, from DuPont's letter to Goodyear in January 1977, that ortho-toluidine "may cause tumors in certain laboratory animals" but found the relevance for humans unconvincing. Goodyear had a computerized system for tracking occupational diseases, Johnson testified, but it didn't work for the Niagara Falls plant because the employees' health insurer wouldn't produce data in a usable form. He said he "may have heard" of the 1970 Soviet worker study. "But just on the basis of these one or two reports, you can't jump to a conclusion and say, 'Aha, that is it.' This simply says the cases occurred in these workers. It doesn't say ortho-toluidine caused the cancer." Even after seeing Halford's 1981 letter about the four stricken workers, the doctor said, "I do not think I was convinced, in my own mind, based on the information I knew about, that [ortho-toluidine] was in fact a human carcinogen." Johnson did obtain monitoring data and medical records from the plant and initiated a phone call with a urologist who was treating three of the workers. But he did not convey a sense of urgency in finding the cause of the outbreak. "The only opinion that I could reach was that it was still a question," Johnson said. This laissez-faire attitude lingered at Goodyear for all of the 1980s and the first half of the 1990s; improvements to processes in Department 245 were made only around the edges.

By the time Halford wrote his letter, Hank Schiro's cancer had migrated from his bladder to his prostate. Both organs were removed—as was his urethra, which had been irreparably damaged by all the cystoscopies—at New York's Memorial Sloan-Kettering Cancer Center, where he spent twenty-three days. The head of Schiro's penis was closed with sutures; he had to urinate through a hole in his body called a stoma, to which a plastic pouch was attached. When the pouch was full, he'd empty it. He had a penile implant, hoping to revive his sex life, but the outcome was unsatisfying. "Because my urethra [had been] removed, my penis would not get rigid in an upright position. It would be in a bowed position," he testified in a deposition. Sometimes the urine-collection bag would come loose and leak. Schiro found this particularly embarrassing during arbitration or contract negotiations: "I would develop a leaker and I would wet my pants and I'd have to excuse myself from the table and go clean myself up."

Bob Bailey's career at Goodyear roughly paralleled Schiro's. Bailey started as a chemical operator in Department 245 in 1957, the year ortho-toluidine was introduced. "I was in that deadly stuff," he remembered. "The fumes were outrageous. People were turning blue all the time." The workers would retch while cleaning the Sparkler filters, gasp while unloading tank cars. When they complained to their bosses, they were assured the chemical they knew as Dominic wouldn't hurt them. For a time, Bailey, who retired in 1982, would de-grease his hands before lunch by dipping them in buckets of reactor waste. "We didn't know no better," he says. "But they [the bosses] were chemical engineers. They *had* to know."

Christine Oliver went on to become co-director of occupational and environmental medicine at Massachusetts General Hospital in Boston, a position she held for thirty-seven years. Despite his own condition, Schiro "wasn't that interested" in bladder cancer when Oliver first came to the Goodyear plant, she recalled, because "that wasn't what I'd been asked to investigate." Her youthful energy and curiosity—and her ghost-writing of Halford's letter to plant manager Pearson—may have kept a bad situation from becoming worse.

CHAPTER 12

REAGAN

I N OCTOBER 1981, seven months after Rod Halford wrote his letter to Goodyear, Steve Wodka left the OCAW. Tony Mazzocchi, to whom he was
inextricably fused, had run for union president in August 1979 and lost
by less than 1 percent of the vote to Bob Goss, described by Mazzocchi
biographer Les Leopold as "the defender of nuts-and-bolts unionism
against Mazzocchi's left-wing militancy." Afterward, Wodka was viewed
with suspicion. He was ordered to vacate his Washington office and
work out of his house in Maryland. He no longer had carte blanche to
go into the field; all his travel required prior approval. Mazzocchi again
ran for president in August 1981 and again fell to Goss by the thinnest
of margins. Wodka was told he'd be reassigned from Washington to an
outpost like North Dakota. His clout was gone. Evoking Jerry Cohen,
his old boss at the United Farm Workers, he thought about becoming
a lawyer, reasoning that it would put him in a better position to help
sickly workers and punish their corporate overlords.

As Wodka parted ways with the union, OSHA was in upheaval. Ronald Reagan had come into office promising to "relieve labor and business
of burdensome, unnecessary regulations." Eula Bingham was replaced
by Thorne Auchter, who owned a construction company in Jacksonville,
Florida, and shared Reagan's belief that overregulation was stifling
American enterprise. Auchter outlined his philosophy in congressional
testimony: "An adversary environment has characterized the past OSHA
program. . . . Indeed, our ability to build a more cooperative relationship

[with business] will depend largely on the professional manner in which the agency's compliance officers conduct themselves. They are to give appropriate attention to matters of dress, conduct and comportment and to use the time of employers and employees wisely. They are to encourage dialogue, offering suggestions for abating workplace hazards."

Industry began to routinely challenge OSHA health standards, tying some up in court for years. Under Reagan's Executive Order 12291, issued in February 1981, the Office of Information and Regulatory Affairs in the White House Office of Management and Budget became a purgatory for new rules and a hangout for industry lobbyists. Auchter scrapped Bingham's ambitious cancer policy and undid a series of "midnight regulations" issued by the Carter administration in its waning days. One was a rule mandating that workers receive written information about the chemicals they used or made—the so-called Hazard Communication Standard. (OSHA reissued the standard in 1983 after industry leaders complained that states were filling the void by passing their own rules; some went beyond what the federal government had planned to do.) Auchter suspended distribution of the gritty Bingham films, which he deemed too political. He denied a petition from Nader's Health Research Group and the American Federation of State, County and Municipal Employees seeking an emergency temporary standard for ethylene oxide, a carcinogenic gas mainly used by health-care workers to sterilize instruments. (In 1983, a federal judge criticized OSHA for failing to act on the chemical, saying it had made "a clear error of judgment" that put some 140,000 workers at risk.)

In its January 1982 report to the Chemical Manufacturers Association's board of directors, the trade group's Occupational Safety and Health Committee offered a glowing appraisal of the new approach: "We have a unique opportunity here to revise and put in place some rules and regulations that we can live with rather than coast these next three years only to be blitzed vindictively again in the future."

One of Auchter's underlings tried to fire Peter Infante, the frizzy-haired epidemiologist who'd done the birth-defects and Pliofilm worker studies in Ohio and was then head of OSHA's Office of Carcinogen Identification and Classification. Infante had maintained publicly and in official communications that formaldehyde, used in building products and embalming, caused cancer in lab animals. This was consistent with a bulletin he had helped prepare and OSHA had issued during the last days of the Bingham era, but it ran counter to the Reagan administration's position. So, when Infante testified at an April 1981 hearing before the Consumer Product Safety Commission, which was considering a ban on formaldehyde-based home insulation, he marked himself as a pot-stirring holdover from the Carter regime. On June 29, Infante received a letter of proposed termination from Bailus Walker, OSHA's director of health standards. Attached to Walker's letter was a letter of complaint from the Formaldehyde Institute, a lobby group.

Infante was spared after then congressman Al Gore Jr. intervened. In a July 1 letter to Auchter, Gore wrote that "the evidence does not justify the removal of this civil servant from office." The congressman added that "a strong case can be made that your agency's action is politically motivated." At a hearing fifteen days later, Gore called the episode "a blatant attempt to rid the government of a competent scientist who happened not to agree with an industry whose profits are at stake." Infante left OSHA in 2002 to consult for plaintiffs in toxic-exposure lawsuits. His warnings on formaldehyde were borne out in 1982, when the International Agency for Research on Cancer—IARC, part of the World Health Organization—declared it a likely animal carcinogen, and again in 2006, when the agency deemed it a human carcinogen.

At a press conference on March 27, 1981, three days before Reagan would be shot and nearly killed outside the Washington Hilton by John Hinckley Jr., Auchter explained why OSHA had decided to reopen the hard-won cotton-dust standard: like all rules, he said, it should be subject to cost-benefit analysis. The reporters seemed less interested in his blow-by-blow account of the solicitor general's legal moves than in his recision of Bingham's heart-tugging cotton-dust brochure. Why, one reporter wanted to know, had he recalled thirty-nine thousand copies? It was biased against employers, Auchter replied. How so? It made a "dramatic statement" about brown lung, Auchter said, in violation of his vow "to disseminate objective safety and health information." He tried to return to the court case but was diverted by yet another question: What part of the brochure reflected bias? The image of Louis Harrell on its cover? "All," Auchter said tartly.

The backlash against the planned gutting of the cotton-dust rule—substituting respirator use for more protective but also more costly engineering controls—was fierce. On May 4, 1981, brown-lung victims, leaders of the Amalgamated Clothing and Textile Workers Union, and AFL-CIO president Lane Kirkland held a rally outside the Labor Department. Press packets distributed before the event—the president of the textile workers' union accused OSHA of "tearing out the very heart of the standard"— caused jitters in the White House, which was trying to quash the narrative that Reagan was anti-worker. The following month, the Supreme Court upheld the standard, saying workers should be fully protected from toxic substances and rejecting the administration's argument that cost-benefit analyses should precede the tightening of regulations.

Auchter was undeterred. Four months after the ruling, OSHA adopted a policy exempting almost 75 percent of US manufacturers from regular inspections. The move drew "cheers from business and brickbats from labor," according to U.S. News & World Report. In 1982, Auchter loosened record-keeping requirements for some 474,000 firms employing more than eighteen million workers. A critical report from Nader's Health Research Group in November of that year drew ridicule from Auchter, who told the Washington Post that the document was "hogwash" and could best be used to "help me start my fire" in the coming winter. By the end of 1983, Nader was calling for the OSHA chief's resignation. "By limiting OSHA inspectors' enforcement efforts he is giving many companies a green light to violate the Occupational Safety and Health Act," Nader wrote in a letter to Reagan. "Workers continue to be overexposed to toxic substances, including known carcinogens, as a result of Mr. Auchter's refusal to issue health standards in a timely and prudent fashion." OSHA's enforcement staff was cut 22 percent during Auchter's tenure, which ended when he took a job with a Kansas construction company in March 1984—one he had absolved of a $12,680 penalty three years earlier. He denied charges of a quid pro quo.

After leaving the OCAW, Wodka went to work as a Washington-based aide to Fred Baron, a pioneer in asbestos litigation, and got a taste of the legal drama that was unfolding in Niagara Falls. In 1985, Baron filed the first lawsuit against DuPont, the main supplier of ortho-toluidine to Goodyear, on behalf of two workers with bladder cancer, Hank Schiro and Charlie Carson. The case garnered no publicity and was settled out of court; Wodka, attending law school at night, was involved on the periphery, helping with depositions. He passed the bar in 1986 and took a job with the venerable Levinson Axelrod personal-injury firm in Edison, New Jersey. Wodka spent about fifteen months at the firm, learning how to break down a complex asbestos- or chemical-exposure case into pieces a judge or jury could easily grasp. He was in no hurry to leave, but Levinson Axelrod's fee structure was such that Wodka, a novice lawyer, was pulling in more money than some of the older partners could abide. Wodka had signed his first Goodyear client, Richard Sullivan, in December 1987, and took the case with him when he went out on his own on New Year's Day 1988, setting up shop in Little Silver, New Jersey, a prosperous town of some 5,700 people an hour south of New York City. He opened his office in a three-bedroom, rented house at 54 Little Silver Point Road. His wife of four months, Ilene, who had worked for Planned Parenthood, had to teach him how to use the word-processing software WordPerfect; he'd never operated a personal computer. Ilene became his courtroom assistant, running at first a projector and then a computer to display documents. She helped him pick jurors during voir dire, slipping him notes advising which ones he should keep. He would win all four of his jury trials; another two hundred or so of his cases would be settled out of court.

Under the New York State workers' compensation statute, Sullivan and other Goodyear cancer victims were barred from suing their employer; in exchange, they'd gain access to a no-fault system that provides wage replacement and lifetime medical care for work-related injuries and illnesses. In fact, this trade-off has shortchanged workers nationwide for a century. Benefits can be unfairly denied—or, if granted, shockingly insufficient. To win meaningful damages, the Goodyear men needed to go after a third-party supplier—in this case DuPont and two other manufacturers of ortho-toluidine. They needed to prove that the suppliers had known of the chemical's carcinogenicity but hadn't adequately warned Goodyear and, by extension, its workers. Wodka suspected that Sullivan was a harbinger of many more cancers from the Goodyear plant.

His case took the course of all product-liability lawsuits. A complaint is filed, the defendant answers, and the discovery process begins. The plaintiff demands to see all documents that may indicate what the defendant knew about the matter being litigated, and when. Depositions are taken, medical and scientific experts' reports are exchanged. The case is set for trial, but few lawsuits ever get that far. A mesothelioma case, for example, might be settled immediately after the plaintiff is deposed if exposure to asbestos is confirmed; such is the ironclad connection between the mineral and the disease. Other cases go on longer but never are heard by a jury; the amount of the settlement becomes the main point of contention.

Sullivan, a big, stoic man, was deposed in Buffalo over two days in April of 1989. Then fifty-five, he was still shaken by his cancer diagnosis almost three years earlier. His testimony followed a pattern that would be repeated as Wodka acquired cases against DuPont. Sullivan was asked to describe, in excruciating detail, the jobs he did at Goodyear and estimate how many times, and in what ways, he might have been exposed to ortho-toluidine. DuPont's lawyer, Paul Jones, quizzed Sullivan on his personal habits, fishing for anything other than the suspect chemical that might have set off the plaintiff's cancer. Did Sullivan smoke? (Yes, a pack or a half-pack of unfiltered Chesterfields a day; he'd quit "off and on.") Drink alcohol? (Not anymore; used to when "socializing.") Drink coffee? (Four or five cups a day.) Consume sugar or artificial sweeteners? (Yes; no.) Eat red meat? (Not often.) Did his home have formaldehyde insulation? (No.) City water? (Yes.) And on it went.

Like Ray Kline, Sullivan grew up in Pennsylvania coal country, in a town called Patton. His father was a coal miner who developed black lung. Sullivan quit high school after his freshman year and went to work at a shirt factory, then for the Altoona Railroad and at steel mills in Johnstown and Bethlehem. He joined the Air Force in 1951, served four years, mostly in supply, and after being discharged headed to Niagara Falls, where he found work at National Carbon (later Union Carbide),

Carborundum, Kimberly-Clark, and Olin Mathieson before landing at Goodyear in 1961.

Sullivan began as an operator in Department 245, charging the Nailax reactors—filling them with prescribed amounts of chemicals and subjecting the mixture to heat and agitation. During the two- to three-hour charging process, fumes from a manhole cut into the top of the vessel were "venting out into the atmosphere right close to you," Sullivan testified. The reaction itself took about four hours.

Nailax is made in batches. The main liquid ingredients, ortho-toluidine and aniline, arrive at Goodyear in rail tank cars, each typically holding between sixty thousand and eighty thousand pounds. The chemicals are transferred by hose to storage tanks. When it's time to make a batch, they are pumped into a premix tank and combined with hydroquinone, a powder the workers knew as Dedic, before being sent to the reactor. A catalyst, ferric chloride, is added to start the reaction. When it's complete the excess reactants are sucked by vacuum to a recovery system so they can be reused. The batch itself, a molten liquid, is sent to a degasser and then a holding tank. It's cooled until it turns into a film that can be sliced into flakes of Nailax, which are carried by conveyor into the bagging area. The finished product contains unreacted ortho-toluidine, to which workers can be exposed. Before the 1990s, there were many other opportunities for exposure. In 1988, NIOSH reported that one manager in Department 245 had "the highest post-shift urine ortho-toluidine value" of any worker surveyed because he was constantly moving around and sometimes stood in for operators. Maintenance workers, known as millwrights, had "greater air and skin exposure potential" than production workers, again because they were so mobile, NIOSH found. And those unfortunate enough to have to clean and replace the Sparkler filters were subjected to hot vapors, rich with ortho-toluidine, for sixty to ninety minutes at a time.

Sometimes, plugged lines had to be steamed open and "the bad stuff you just let go on the floor," Sullivan said. "That stuff laid in the building forever. You washed it down as much as you could, but you don't get it all off." The floor drain was stopped up "almost constantly."

The ventilation system was unreliable; it took up some of the fumes but "missed most of them." Sullivan remembered seeing "little heat waves, and then you had dust. You had it all." Sometimes a worker would inhale too much of this floating mixture and be told by a buddy, "Hey, you're turning blue." In the 1960s, this was treated more as a curiosity than a dire threat to one's health.

In the early 1970s, Sullivan moved from Department 245 to maintenance. A millwright, he was dispatched throughout the plant to fix bad valves or pumps and continued to be exposed to ortho-toluidine. One of his jobs was to hook up hoses to the bottoms of tank cars loaded with the chemical, which was "thick like oil" and "smelled funny." Sometimes there were spills, and he got the liquid on his hands, arms, or legs. Sullivan also was expected to climb railcar manways, open a hatch on top of the car and dip in a cup attached to a four-foot stick to grab samples of the load. The samples were sent to the lab, where they were tested for purity; the unloading of ortho-toluidine from the car to the tank farm took place only after the bosses gave the go-ahead. The chemical was pumped in batches from the tank farm—where Sullivan's face once was sprayed and scarred after a line rupture—to Building 32, which housed the Nailax reactors.

Sullivan learned he had bladder cancer in June 1986. He'd gone to the doctor after he'd begun passing blood while urinating, and because swelling in his foot and leg would not go away. The doctor found a tumor and removed it that month. Sullivan underwent weeks of chemotherapy and afterward had a cystoscopy every three months. At the time of his deposition, there was no sign the cancer had returned. "I don't ask for any more than that," Sullivan said.

Jones, the DuPont lawyer, asked Sullivan whether his activities had been restricted by the disease. Yes, Sullivan said, but he wouldn't be specific. He was still working at Goodyear and simply wanted to retire and take care of his family. He urinated frequently. He got "moody. I don't want to associate with nobody." He was terrified of a recurrence.

JONES: "Anything else?"
SULLIVAN: "That's enough."

By the time Sullivan and DuPont settled under confidential terms in 1990, Hank Schiro had been dead nearly four years. Toward the end of his life, Schiro had endured a string of health setbacks, and treatments that seemed to have been designed by sadists. In October 1984, a tumor was found on his urethra; his doctor had detected it using a technique called a looposcopy, in which he inserted a foot-long tube through Schiro's stoma to see if there was any blockage. There was. Schiro was conscious during the procedure, which, he said in a deposition, caused "excruciating pain, like somebody sticking a knife in you and pulling it out." This took fifteen to twenty minutes, after which the doctor determined that Schiro required surgery; the tumor was interfering with his kidney function. The doctor removed the tumor and rebuilt Schiro's ureter—the duct through which urine passes from the kidney to the bladder—turning it from straight to bowed. Schiro spent ten days in the hospital.

Schiro's condition stabilized until August 1985, when he passed blood immediately before leaving on a vacation to San Francisco. His doctor did another looposcopy and found an "extensive mass" on the left side of his kidney next to the ureter. Schiro underwent a nephrostomy—the drilling of a hole in his back to try to get the kidney functioning again. This, Schiro said, took about twenty-five minutes and felt "like somebody sticking you with needles constantly." The doctor inserted a tube in the hole and left it for four days; blood drained steadily from it. When the doctor went in to reconstruct Schiro's ureter he found a liver malignancy. Schiro's kidney was removed, and "I now have cancer of the liver." He had just finished a round of chemo the day before his deposition on November 4, 1985. He was nauseous. He was told his hair would fall out—another indignity to add to his impotence. He hadn't slept for two nights because of pain on his right side and persistent burping and hiccupping. His wife had stayed up with him. His prognosis was "not very hopeful," he admitted, and, indeed, he would live only eleven months longer.

Wodka, though outwardly businesslike, felt for his suffering clients. What drove him most of all, however, was what he perceived to be untenable, mean-spirited behavior by defense counsel. Many plaintiff's lawyers aren't built for such conflict and resort to transactional work—representing auto-accident victims, for example. Wodka's twelve years with the OCAW prepared him for legal combat: every time he landed a punch, he expected a counterpunch.

CHAPTER 13

HARRY MOVES UP

IN 1986, THE YEAR HANK SCHIRO DIED, Harry and Diane Weist were living in a 14-by-17-foot trailer in the Sunny Acres mobile-home park on Niagara Falls Boulevard. It was a nice enough place, but they had three young children—Holly, born in 1981, Harry Jr., born in 1983, and Kristan, born in 1985—who needed more room. Diane, who, in her words, could be "bullheaded," decided that she and Harry should buy a small farm. She found one she liked: a thirteen-acre plot in Youngstown with a decaying, 148-year-old house on it. Harry was opposed to the purchase, but Diane sulked for two weeks and wore him down. He assumed, correctly, that the farm would drain their finances and occupy almost every moment of their free time. The house had no furnace, only a wood-burning stove in disrepair. Harry hit his head on the pipe one day, knocking it loose from the wall. The house filled with smoke.

Even Harry agreed, however, that the farm was good for the children. There were horses, cows, pigs, chickens, and ducks. The family grew corn, tomatoes, cucumbers, squash, and grapes. Diane was fit then and could carry hundred-pound bags of grain for the animals. Harry was pulling in lots of overtime by working brutal, "continuous" shifts: seven consecutive days of 3-to-11's, followed by two days off; then six days of 7-to-3's, followed by one day off; and, finally, seven days on the midnight shift—11 to 7—which he despised, followed by five days off. The cycle started over after that. The job suited Harry because he worked alone in the tank farm, beyond the grasp of most of the bosses. They had other things to do and didn't bother him if he got his work done. The exception was a micromanager who felt compelled to hand out assignments. An infuriated Harry once whacked him in the shin with a shovel during an argument. Harry managed to keep his job, and the two later became friends.

In January 1988, Harry finally got out of production at Goodyear and the erratic shift work that came with it. His best friend, Robert Dutton, had been nagging him to move to the maintenance department, which offered better pay, daytime hours, and plentiful overtime. Harry, who didn't enjoy fixing things, wavered, but Dutton, who did, egged him on. After practicing off shift, they passed their tests and became Class B millwrights. Their jobs—unclogging pipes, fixing pumps, seals, and agitators—took them to every corner of the plant: Department 145 (vinyl), Department 245 (rubber chemicals), the warehouse, the tank farm. Their boss was Ray Kline, Harry's father-in-law, which made for some awkwardness early on. For a time, Ray assigned Harry the worst jobs to avoid the appearance of favoritism; eventually Ray eased up, and the two men came to relish their impromptu discussions about family in Ray's office. When Ray did something to irritate him, Harry would threaten to deny access to the grandchildren. Ray found this prospect unbearable and would appeal to his wife, Dottie, who in turn would appeal to their daughter, Diane, who would tell Harry to knock it off.

Within a year of moving over to maintenance, Harry and Dutton had earned their Class A certificates, making them eligible for higher pay. They became proficient in their craft and were kept busy repairing equipment, unplugging lines, and hooking up hoses to tank cars. When the mood for indolence struck, they hid in the locker room or the tank farm. Ray, having done the same thing when he was younger, usually found them and sent them back to work. He was so attentive that the men developed a signal; an extended arm held low to the ground meant, "Have you seen the Little Guy?" This gesture—a reference to Ray's compact stature—was met with either a nod or a headshake. Ray, known as "One-Liner Kliner" for his droll wit, managed to be both unfailingly loyal to Goodyear and respected by his underlings.

Harry's move to maintenance from Department 145 did not spare him from exposure to ortho-toluidine. In fact, it increased. When he was on his hands and knees connecting a hose to a tank car, for example, residual amounts of the chemical would spill onto his clothing and boots and soak through his cotton gloves. When he'd pull apart a pump containing reactor waste known as "recycle," the liquid would go everywhere. It pooled in a pit beneath the pump and squirted from the pump head. When it puddled on the floor, workers used Nailax flakes as an absorbent, creating a mud-like, hypertoxic solid. "They never cleaned it up," Harry later testified. "It was always, you know, I hate to say . . . a shithole back there." Workers sometimes used hoses to force spilled recycle into open sewers, where iron residue from the repugnant Sparkler filters also would be dumped.

As Harry was becoming acclimated to the maintenance shop, bladder cancer was advancing through the Goodyear plant. He was aware of this development but didn't see it as cause for concern. Wodka, on the other hand, did. At a dinner meeting in New York City in November 1987, he told OCAW vice president Robert Wages that the number of cases had risen to eight. Wodka wanted the union to get behind an investigation; Wages told him to put the request in writing. Six days later he did, arguing that "someone needs to ring the alarm concerning ortho-toluidine. . . . The key issue here is that OSHA does not regulate ortho-toluidine as if it were a human bladder carcinogen. Second, there is a serious question whether workers who are currently exposed to ortho-toluidine are adequately warned about the bladder cancer risk and are adequately instructed how to protect themselves." In December, Wages asked both the EPA and OSHA to investigate. The EPA did nothing. OSHA took air samples during a plant inspection in February 1988 but issued no citations because the exposure limit for ortho-toluidine wasn't exceeded.

Only NIOSH came through. At Wages's request, Sylvia Krekel, the union's health and safety director, wrote to William Halperin, chief of the agency's Industry-Wide Studies Branch, listing the names and dates of diagnosis for the eight victims and asking for a health hazard evaluation at the plant. NIOSH investigators did a walk-through in early May and returned in August to collect the personnel records they needed to do a cancer-incidence study. During the walk-through, the investigators noted with concern the cleaning of the Sparkler filters. "During the cleaning procedure there were visible vapors coming off the filters and the workers wore no respiratory protection," they wrote in an interim report issued in 1989. While air levels of ortho-toluidine and other chemicals of interest were far below legal limits, "there was an obvious potential for inhalation, as well as skin exposure at the neck and face of the operators. . . . Biomonitoring or skin exposure assessment of these workers has not been conducted by Goodyear." The company had been collecting air-monitoring data since 1976, but the readings provided a false sense of security. All were under the limits by a comfortable margin.

The NIOSH investigators' findings, published in the *Journal of the National Cancer Institute* in 1991, shook even the most hardened union men: there were thirteen confirmed cases of bladder cancer at the Goodyear plant, about four times what would be expected in the general population of New York State, excluding New York City. Two other cases had been left out of the NIOSH analysis, one because it was believed, incorrectly, to have occurred before the study period began, the other because it was found after the study period had ended. This meant that the total number of cases at the plant was fifteen. Seven cases included in the study had occurred among workers assigned to Department 245, where ortho-toluidine and another aromatic amine, aniline, were used. The incidence rate there was elevated more than sixfold, with an average latency period of twenty-three years. Given that ortho-toluidine had been found to cause bladder tumors in rats, it was likely responsible for the excess, the authors wrote, "although aniline may have played a role." Significant levels of both substances had been found in exposed workers' urine post-shift "despite low air concentrations," suggesting Goodyear's emphasis on inhalation as opposed to absorption had been misguided. The cotton gloves and street clothes Harry Weist and his colleagues wore during the messy Nailax production process were all but useless as barriers to the chemical. Goodyear began making its employees wear impervious gloves, coveralls, and boots and supplied-air respirators when cleaning the filters or performing other high-exposure tasks. It welded pipes to prevent leaks, improved ventilation, and automated product sampling from tank cars. All of this could have been done years earlier, of course, but Goodyear—either oblivious, recalcitrant, or both—needed steady prodding from the government and the union.

DuPont was hardly blameless. It had been encasing its aromatic-amine production workers at Chambers Works in moon suits since the mid-1950s but had been in no hurry to warn Goodyear or other customers about the pitfalls of getting ortho-toluidine on one's skin. It knew that the only accurate way to measure exposure was to test the urine, and that the urine of workers wearing Chem-Proof suits was uncontaminated with ortho-toluidine. This was not the case for those who wore regular clothing. "Air analysis alone, even on a personnel monitor basis, is not enough to guarantee freedom from occupational illness," Adrian Linch, then an industrial hygienist at Chambers Works, wrote in 1974. But none of the chemical-safety fliers—known as "material safety data sheets"—DuPont issued to Goodyear mentioned the need to conduct routine urine sampling among exposed workers. Nor did DuPont advise Goodyear that the OSHA exposure limit for ortho-toluidine—5 parts per million, measured in air—was formulated not to protect workers from cancer but to guard against the chemical's immediate, toxic effects. As a result, Goodyear did only air monitoring in Niagara Falls until 1992, congratulating itself on the consistently minuscule levels of ortho-toluidine it found and conveying the good news to workers. "We relied on the OSHA standards," Goodyear's corporate manager of industrial hygiene, Joseph Holtshouser, testified in a deposition. "And if there was any new information that came from the suppliers, we relied on that."

DuPont stopped making ortho-toluidine in 1995, leaving First Chemical Corporation of Pascagoula, Mississippi, as the sole American manufacturer of the liquid. Like DuPont, First Chemical understated the pitfalls posed by ortho-toluidine, falsely claiming that an alert issued by NIOSH after its Niagara Falls investigation "does not say that o-toluidine and aniline cause cancer in humans." When First Chemical jettisoned the product in 2009, Goodyear found suppliers abroad. Nailax—sold as Wingstay 100—was too important a product to abandon.

CHAPTER 14

WHAT IS BLADDER CANCER?

B LADDER CANCER—not the most lethal of tumors but perhaps the most unpredictable—takes about seventeen thousand lives in the United States each year, disproportionately striking men. Because the disease is at the crux of the Goodyear story, it's worth pausing here to consider what it does to the human body. One of the best summaries—a 162-page patient's guide—was produced by Dr. Khurshid Guru, who chairs the Department of Urology at the Roswell Park Comprehensive Cancer Center in Buffalo. In the book, Guru, who has seen fifty to a hundred bladder-cancer patients a year since 2005, walks the shell-shocked victim through detection, treatment, and post-operative care. The first chapter is appropriately titled "Why Is This Happening to Me?" The last is called, ominously, "If My Options Seem Limited." The twenty-four chapters in between are full of practical information on such topics as the side effects of chemotherapy; sexual dysfunction after surgery or radiation; and the proper care of one's surgically installed "neobladder." The graphic that takes up pages 40 and 41 is especially disquieting. It shows how bladder cancer—survivable if caught early, ghastly if un-contained—can evolve from pathological classification Ta (the tumor is confined to the bladder's innermost layer) to T4 (the tumor invades the prostate, uterus, vagina, or pelvic or abdominal wall). In the drawing, Ta resembles a small cluster of berries affixed to the bladder lining, T4 an unstoppable blob that has breached the bladder wall in search of new tissue to plunder.

About eighty-two thousand new cases of bladder cancer are diagnosed annually in the United States, according to the American Cancer Society. The disease, the ACS explains, commences "when cells that make up the urinary bladder start to grow out of control. As more cancer cells develop, they can form a tumor and, with time, spread to other parts of the body." Risk factors include smoking; exposure to chemicals in the workplace; use of the diabetes drug pioglitazone, sold under the name Actos; and consumption of water contaminated with arsenic. White people, according to the ACS, are about twice as likely to develop the disease as Black and Latino people. The risk increases with age.

Few victims illustrate bladder cancer's unbridled ferocity better than Joseph Nicastro. During the last year of his life, Nicastro, a retired chemical worker from New Jersey, was punctured with tubes and way-laid by chemotherapy and radiation treatments. He had bouts of severe diarrhea. His buttocks blistered and cracked. His right leg throbbed from the cancer that had spread from his bladder to his bones. Born in 1943, he'd quit high school in the twelfth grade, worked for a few years at his father's wastepaper-hauling company, and spent six miserable months in the army at Fort Dix—mostly in the brig, by his telling, for hitting another soldier—before landing at Pfister Chemical in Ridgefield, New Jersey, in January 1966. He started out as a laborer, unloading drums of chemicals and shoveling muck in what he later described as a "very dusty" environment, and was promoted to assistant operator on the overnight shift. It was here that he got his first dose of ortho-toluidine, inhaling the vapors and absorbing the liquid through his pores as he charged reactors. In February 1967, with a colicky newborn son and an urgent need for sleep, he quit Pfister—which refused to let him change shifts—to join Patent Chemical in Paterson. For the next thirty-five years the dye factory at 335 McLean Boulevard, next to the Passaic River, would be his province for up to eighty hours a week.

An operator in Building 6, Nicastro made dyes known as Red B, Yellow 8, and Blue 8. Red B, used to color gasoline, was Patent's signature product. It was made in open 5,000-gallon tanks; among its ingredients was ortho-toluidine, pumped from 55-gallon drums and supplied by DuPont. It had a "horrible odor . . . worse than a skunk," Nicastro would testify in a deposition many years later. When muriatic acid was added to one of the reactors, the temperature of the mixture inside would spike. To cool it down, workers threw in 50- to 75-pound chunks of ice. This process created fumes so overpowering that the operators covered their mouths with wet rags. Nicastro called the conditions "primeval."

After Patent sold the plant to Morton International in 1968, hygiene gradually improved but worker exposures persisted. Nicastro left Building 6 for newly opened Building 11. He'd receive batches of Red B and other concoctions from the manufacturing side, dry them and filter them. "The flakes, they used to get all over the building, and you could smell it, you could see it in particles floating in the air," he said. Chemical waste in the sewer system "went straight . . . to the Passaic Valley sewer, no in-between stuff." It got so bad in the building "you had to step outside in order to breathe." There was a "three-foot sewer system with grates across and all you saw was the vapors coming out of these systems and it was horrible, horrible debris." Ventilation was bad here, as it had been in Building 6. Morton tried to fix the problem, repairing fiberglass duct work that had been eaten away by acids and lost suction as a result. After a while, the acids would chew through the ducts again.

The Morton workers voted for OCAW representation in 1970. They began holding safety meetings and in time were issued aprons, goggles, face shields, and what Nicastro called "the right kind of shoes. . . . That is the number-one place where the chemicals are absorbed—through your feet." Nicastro, a shop steward, began wearing a half-face respirator while adding chemicals to the reactors, as required, but ultimately would take it off. "After we finished with the charging," he said, "the company said the levels of everything [were] OK, to remove your mask." Gloves had always been supplied, but the quality was uneven. Some "you could tell were seconds" and were missing patches of neoprene, Nicastro said. "And if the chemical got on the glove, it would saturate and go . . . right into your hand."

In 1999, Morton sold the Paterson plant to chemical giant Rohm and Haas. Nicastro stayed on for another three years, retiring in 2002.

He and his second wife, Pam, spent the next six years living a low-key existence in Ocean Township, New Jersey. Joe cleaned the house, did the grocery shopping, and kept the cars running. Pam worked part-time as a babysitter. They rarely traveled, preferring to go to movies, stroll on the boardwalk, and eat out once a week. Everything changed on May 28, 2008, when Joe, who'd been having trouble urinating, was diagnosed with bladder cancer. He'd been advised by Rohm and Haas that an annual mail-in urine test offered to retirees had found abnormal cells. A cystoscopy revealed a tumor.

The diagnosis set Joe on a course of treatment that bordered on the inhumane. Having sued DuPont and other chemical suppliers the month after his diagnosis, he gave three videotaped depositions. At the first session, in November 2008, Nicastro sits on his living room sofa and crisply answers questions from Steve Wodka, his lawyer. He is sixty-five, with a full head of gray hair, and does not seem especially ill. Unseen in the video are the two nephrostomy tubes that run from his kidneys, through holes in his back, to urine-collection bags affixed to his legs. Wodka asks him to describe any physical symptoms he's experiencing. The cancer, he explains, has moved from his bladder to his bones and lymph nodes. The pain in his right hip is sometimes "excruciating, and if I step the wrong way or something, I'm going down." What else? Wodka asks. "I'm sitting here with you talking and I'm so uncomfortable," Joe says. "I got tubes coming out of my back. I can't get a good night['s] sleep. I can't go anywhere. I feel like I'm a prisoner in this house, that all I can do is watch TV and life is boring, and this is not living."

At the second session, in January 2009, Joe appears thinner and grayer but is still lucid. Wodka asks whether his health has changed in the two months since the first deposition. It has, Nicastro says. Radiation treatments begun in November made him vomit several times a day and gave him diarrhea. At his stepdaughter's wedding, he was upstairs in the master bedroom, next to the bathroom, "and I couldn't control myself, and . . . the diarrhea went all over the floor." Pam was taking pictures downstairs. Joe called her, crying, to tell her what had happened. "It's just very degrading." The radiation treatments had been directed at his buttocks, and the skin had cracked. "And I couldn't even sit down. It was horrible." In December the doctors were able to remove the tube on his right side after installing a stent to connect his kidney to his bladder. The tube on the left side stayed. The stent made it difficult for him to control his urine flow. He once urinated "all over the kitchen floor."

The diarrhea and vomiting had stopped by early December, after he'd finished the radiation treatments. Now he was taking chemo once a week. It gave him hives all over his body. "My hands are so itchy they feel like . . . they're on fire."

By the final session, in February 2010, Joe has turned into an old, bald man with a vacant stare. He is confined to a bed at the Meridian Nursing and Rehabilitation Center in Shrewsbury, New Jersey, and looks much older than his sixty-six years. Wodka asks how he's feeling.

"Horrible," Joe says.

"Do you have any problems?"

"I have a lot of problems."

"What are they?"

"Problems breathing, problems sitting, problems talking—all kinds of problems."

Joe had come to the nursing center two days earlier from Monmouth Medical Center, where he'd been treated for bone cancer in his right leg. He'd received chemo forty or fifty times in the past year. "I get awfully nauseous, sick," he says. "I can't hold anything down. I throw up. I get weak." Over the same period he'd been hospitalized a half-dozen times. The skin was still peeling from his buttocks. At one point Joe deposits a mouthful of phlegm into a cloth held by Pam. "It's that bile," he explains.

Wodka asks Joe if he's planning to go home at some point.

"I hope it's not in a box," Joe says. Two weeks later, he would be dead.

I met Pam at her townhouse three years after Joe died. The last year of his life was "pure hell," she told me. She'd been his full-time caretaker, flushing the tubes that drained his kidneys and emptying the bags that collected his urine. She helped him go to the bathroom, gave him sponge baths, took him to doctor visits, and tried to boost his flagging spirits. "He was so scared. His mind was racing. He would always say he was sorry to me. I would say, 'Why are you sorry? You had to make a living.'"

CHAPTER 15

THE GOODYEAR
EPIDEMIC SPREADS

O N JUNE 11, 1990, NIOSH disclosed the gravity of the bladder-cancer situation at Goodyear in a letter to current and former workers. What for many employees had been an unsettling rumor—a feeling shared in locker rooms, bars, and bowling alleys—was now quite real. The letter, signed by Lawrence Fine, a physician who headed the agency's Division of Surveillance, Hazard Evaluations and Field Studies, and epidemiologist Elizabeth Ward, who'd led the NIOSH study of the plant, explained that ortho-toluidine "has been found to cause cancer in rats and mice and bladder cancer in rats." While NIOSH researchers believed the chemical was mostly responsible for the cancer surge, another ingredient in Wingstay 100, aniline, might have had a role. "We therefore recommended to the company that exposure to both chemicals be reduced to the lowest feasible level," Fine and Ward wrote. Some unnerving context was offered: "Unfortunately, we cannot predict which workers may develop the disease. This is similar to the situation of cigarette smokers. While we know that cigarette smokers as a group are at increased risk for several diseases (including bladder cancer), we also know that not all cigarette smokers get sick." Attached to the letter was a fact sheet outlining symptoms of bladder cancer: blood in the urine, with or without associated pain; "hazy, cloudy or murky" urine; frequent urination; weak flow; a lump on the groin or lower abdomen. Those who worked in Department 245, the sheet said, should "make every effort to avoid contact with [ortho-toluidine and aniline], particularly contact with your skin or work clothing."

After the NIOSH letter went out, Goodyear, no longer able to play down the outbreak, began to make changes at the plant. Processes that had been open to the atmosphere were closed. Ventilation improved. Production workers were hounded to wear protective Saranex suits (made, ironically, by DuPont). Ed Polka, an OCAW safety representative at the time, was one of the designated nags, often arguing with old-timers who resisted suiting up in 90-degree weather. "Get the fuck out of here," they'd tell Polka, to which he'd reply, "Are you a fucking idiot? You need to put the stuff on." Polka, who worked at Goodyear for thirty-eight years, heard no mention of bladder cancer, or ortho-toluidine, when he started bagging Nailax in 1979. It wasn't until he became a safety rep in the early 1980s that he learned the chemical could cause cancer—and do so with ruthless efficiency when people weren't protected. Goodyear was in denial well into the 1980s, Polka said. During one union meeting its corporate medical director at the time, Clifford Johnson, tried to convince the men they had a better chance of getting bladder cancer from grilled meat than from any chemical in the plant.

Dr. Philip Aliotta opened his urology practice in Buffalo in 1987, and before long started seeing patients from Goodyear with advanced forms of the disease. "A number of them went on to die," he told me. Steve Wodka contacted him to see if he thought there might be a connection to the ortho-toluidine exposures. "At first I was hesitant to do anything," Aliotta remembered, "but once I reviewed the [scientific] material and saw what was known and what was kept from the workers, I did all the work pro bono." His posture—that Goodyear should have foreseen the damage ortho-toluidine exposures would inflict on its workforce—put him at odds with one of his former professors: Dr. Robert Huben, chair of urologic oncology at Roswell Park. Huben, hired by Goodyear as an expert witness, insisted that smoking had been the culprit, but Aliotta stood firm. "I felt that [Goodyear] betrayed their workers because they knew what this product did," he said. In return, he was branded a "whore" by doctors aghast at his gall. He found this farcical, given that he wasn't being paid for his services.

Aliotta used a simple analogy to explain the bladder's three layers to his patients: The mucosa was like the carpet in your house, the submucosa the carpet pad. If the tumor was confined to one of these layers, it could be removed if it recurred. Once it reached the muscle—the floorboard—the picture changed radically. The bladder would have to be taken out or, for those who weren't good candidates for surgery, treated with chemo and radiation therapy. Aliotta was deposed in Ray Kline's case against DuPont and was needled by the defense lawyers, who wanted him to speculate on all the possible causes of Kline's tumor and that of a coworker. Aliotta gave it right back to them, noting that they were ignoring the elephant in the room: ortho-toluidine.

Wodka's cases against DuPont were built on epidemiology—the study of the distribution and determinants of disease within groups of people. He had a decided advantage: he could refer to government data showing pronounced excesses of bladder cancer among the Goodyear workers in Niagara Falls. In the study that would be published in the *Journal of the National Cancer Institute*, for example, NIOSH's Ward and her co-investigators reported that the risk of bladder cancer for all workers at the plant was 3.6 times what would have been expected in the general population. For those known to have been exposed to ortho-toluidine, it was 6.48 times the expected number. Relative risks this high are rare, even in studies of workers exposed to occupational carcinogens.

Goodyear wasn't a defendant in Wodka's cases and in theory had little to lose by admitting ortho-toluidine was a hazard. Well into the 1990s, however, even as it upgraded equipment and processes, the company was still challenging the NIOSH conclusion that exposures to the chemical and possibly aniline were likely responsible for the cancer spike. In a 1992 memo to the Niagara Falls plant manager, Goodyear corporate industrial hygiene manager Joseph Holtshouser resisted the idea of regularly testing employees' urine for the chemicals, asserting that "the cancer-causing potential of o-toluidine and aniline in humans has not been demonstrated. Consequently, measuring these chemicals in workers' urine has no value in assessing the risk of workers contracting bladder cancer." Holtshouser also recommended that the company not join the OCAW in pressing OSHA for an emergency temporary standard for ortho-toluidine, saying there was "no credible scientific evidence" to justify such a move. "We support your continuing cooperative efforts to reduce potential for contact with chemicals in the workplace through the implementation of engineering controls, process changes, improvement of personal protection equipment and personal hygiene practices," Holtshouser wrote. "In addition, for those already at risk for the development of bladder cancer, maximum participation in the program for early detection and treatment should be encouraged." While zero exposure to ortho-toluidine was not a possibility, he wrote, Goodyear would "continue its on-going exposure reduction program in order to minimize contact/exposure with chemicals and to provide a successful level of protection for Department 245 employees."

In 1996, Goodyear's corporate medical director, Donald Sherman, suggested edits to a letter NIOSH planned to send to Niagara Falls workers. Sherman worried that the letter would "strike fear in the hearts of our associates," and noted that levels of ortho-toluidine and aniline had fallen significantly since NIOSH first visited the plant. Besides, Sherman wrote to the agency's Ward, there was still no solid evidence the chemicals had caused the cancer outbreak. A "more plausible hypothesis," he argued, was that another substance released in high amounts in the early days of Department 245 but brought under control by 1966—4-aminobiphenyl—was responsible. "We have believed all along that aniline and o-toluidine did not cause the cancers in the Niagara Falls plant," Sherman wrote. Levels of both chemicals in the plant had remained steady from 1966 to the late '70s, and no worker who'd developed cancer had started at Goodyear after 1966; therefore, the problem must have been solved. In fact, at least thirty-five workers with start dates after 1966 would be stricken with bladder cancer.

Because of latency, the course of the epidemic in Niagara Falls had been set decades before Goodyear began to make improvements. Chemicals had been handled carelessly, and the bill, measured in human suffering, had come due. Bladder cancer snared front-office managers as well as operators and maintenance men. It got a female employee and a worker's wife. It even got a college student who spent a few months at the plant. The unlikeliest victim, in retired millwright Robert Dutton's view, was a member of the maintenance department known for his obsession with cleanliness. "When he went to the bathroom," Dutton said, "he'd take a paper towel out, clean his thing, pee, take another paper towel, clean it and put that shit away. Later on we found out he got the ginch. He's gone now."

Rod Halford became case No. 18 after filling his toilet bowl with blood on August 4, 1992. He'd made an appointment with a urologist several weeks earlier, not because he was symptomatic but because Goodyear had informed him it was planning to change retirees' medical coverage, and not for the better. Halford knew bladder cancer had invaded the plant and wanted to have any medical issues addressed quickly so he wouldn't be stuck with a pile of bills. His bloody urination motivated him to move up his appointment, from the 14th of August to the 9th. His bladder tumor was removed the next day as he chatted, unanesthetized, with his doctor.

When Halford was first deposed in his lawsuit against DuPont and other chemical suppliers, in May of 1993, his cystoscopies had come back clean. He'd avoided chemo and radiation. But he no longer felt comfortable hunting and fishing in New York's Southern Tier, near the Pennsylvania border, fearing he'd need medical attention deep in the woods. He also steered clear of Buffalo Bills games, which he used to attend four times a year, not wanting to scale the stairs of Rich Stadium with his unpredictable bladder. He'd been in the thick of the investigation into heart disease at Goodyear more than a decade earlier and had followed the cancer cases as they mounted: Schiro, Carson, Sullivan. Still, he'd been surprised when the ginch found him.

Case No. 19 was Dick Prato, who started at Goodyear as a janitor in Department 245 in January 1963, became a production operator in March of that year, and worked as a utility man and chemical operator until July 1994. He'd performed unpalatable tasks, including bagging Nailax, a process that left dust pasted on his arms and face and stained his clothes. He'd cleaned the much-loathed Sparkler filter, coating his skin with the dregs of the dark brown Nailax mixture. He'd unloaded tank cars of ortho-toluidine and cleaned the insides of reactors. He'd worn leather boots while doing these jobs, and the liquid soaked through to his socks. He'd wash his feet with soap and water, though not always immediately, and didn't always change his wet clothing. The odor was off-putting, but he wasn't worried; he trusted his bosses to tell him what he needed to know.

Though he'd become vice president of Local 8–277 in 1985, Prato hadn't realized there was a cancer problem at the plant until he attended a meeting at the union hall on January 23, 1990. Ward, of NIOSH, and Wodka both spoke that evening. Five and a half years later, after urinating blood during a camping trip, Prato got the diagnosis he feared. The tumor was removed, and he underwent six weeks of chemo—"not a pleasant procedure," he said in his deposition. "They insert a catheter through the penis into the bladder, inject the chemo treatment and remove the catheter. You keep it there for two hours. Urinate it out. And then go back the following week for another." The side effects? "Oh, incontinence, discharge, cramping at times, very bad burning during urination." The illness and treatments had ruined his sex life, dissuaded him from taking long motorcycle rides to the Adirondacks, led to a job reassignment and pay cut, and made him anxious—too nervous to eat at times. "I worry almost constantly," he told a lawyer. About what? "About it returning and being worse and moving to someplace else."

The cancer did come back in 1997 and 2007, and Prato rode out two more rounds of chemotherapy. The side effects were like an acute bout of the flu: he would alternately freeze and sweat. The disease's recurrence scared Prato into scheduling more frequent cystoscopies, the aim being to keep the cancer from breaching the bladder wall. "Once it goes through the wall," Prato told me, "it's going somewhere else—your spine, your scrotum."

Uncounted among the Goodyear cases was the bladder cancer that attacked Dorothy Kowalski, wife of electrician and supervisor Lou Kowalski. Lou had begun working in maintenance at the plant in 1958 and been sent to Department 245 many times. For twenty-nine years Dorothy washed Lou's clothes, including his reeking work gear: T-shirts, socks, underwear, shirts, pants, jackets, towels. On Fridays Lou would bring his Goodyear laundry home in a cotton duffle bag and throw it down the basement stairs. Dorothy would sort it and wash it on Saturday mornings. It smelled like "chemical plants," she testified in a 1995 deposition. When Lou came home for lunch, once or twice a week, Dorothy made him leave his stinking jacket out on the landing. He left a yellow stain on the pillowcases and a yellow ring on the collars of his chambray shirts. He sullied the white brick floor of the house with his dirty shoes.

Dorothy, who had a history of urinary tract infections, gave what she believed to be a routine urine sample at her doctor's request in December 1983. The following month she learned that blood had been found; a cystoscopy performed under local anesthesia a few days later unearthed a bladder tumor one milligram in length, which was removed. She'd had cystos periodically after that, and the cancer hadn't returned. But the anxiety was omnipresent. "Every time I go in for a cystoscopy," she explained in her deposition, "I get all upset that it's going to come back and all that. . . . Every time I get an infection, I say, 'Oh, my gosh. Is it coming?' . . . Every time I have to go to the bathroom quick—about four, five times a night—every time I get pains in my stomach or something, I say, 'Oh, geez,' then they go away, but in the meantime I'm worried that there's something developing." Before each cysto the doctor would pat her reassuringly on the arm.

Dorothy and Lou had sued Goodyear—they weren't bound by the workers' comp bar because Dorothy wasn't a Goodyear employee—alleging her disease was the product of secondhand exposure to ortho-toluidine. During the deposition, Goodyear's lawyer, Diane Bosse, had Dorothy review the most mundane aspects of her life: where she'd lived, whether she'd consumed municipal water or hot dogs, smoked cigarettes, or used a wood-burning stove. Dorothy was asked to list her jobs (garment, paper factories) and hobbies (sewing, ceramics). Bosse was fishing for anything to divert attention from her client's carcinogenic product.

Two months after his wife was deposed, Lou Kowalski gave his own testimony and recounted some of the ways he'd been exposed to ortho-toluidine. He'd worked in the tank farm, taping valves that leaked the chemical and soaked his clothing ("That stuff didn't dry," he allowed.) He'd repaired or changed out pump motors, working on his knees, his clothing getting "extremely damp, wet. In other words, it would become uncomfortable." There had been more regular exposures in the area of the Nailax flaker. "When the hot Nailax came down out of the reactor, or holding tank, it was steaming," Lou testified. The vapor was rich with unreacted chemicals, including ortho-toluidine. "We also got splashed with the liquid Nailax on our clothes," Lou said, leaving black, sticky spots ranging in size from the head of a pin to a dollar bill. Had he rinsed it off? Bosse asked. "You don't rinse Nailax off," Lou said.

Years later, Harry Weist, Ed Polka, and Robert Dutton—known to his friends as Dickie—tried to recall when they began to suspect something was awry at Goodyear. Dutton said he heard "rumblings" in the early 1980s that ortho-toluidine "was bad for you." Around the same time, a federal law began requiring chemical manufacturers to maintain material safety data sheets to warn workers about the hazards of their products, but Polka said he and most other workers didn't know to ask for them. Even if they had, they'd have encountered technical documents so opaquely written that they would have come away no more enlightened than when they started, and possibly less.

DuPont's name came up, triggering a torrent of vitriol. "DuPont should be beat on the head with a club, because they knew," Polka said, referring to the company's circumspect handling of ortho-toluidine at its own facilities in the 1950s and its failure to alert Goodyear to the chemical's cancer-causing propensities for two decades. Even after DuPont had delivered its mild warning to Goodyear in 1977, word didn't filter down to the workers in Niagara Falls. "They didn't tell Dickie and Harry and me and the guys I worked with, 'Say, you know the product we're making? We got guys getting bladder cancer, and you can get it,'" Polka said. That, he said, was why DuPont was "bailing water out of a fucking rowboat with a hole in it"—facing a stream of lawsuits from Goodyear workers.

As bad as the epidemic had been, it would have been worse had the union not asked for the NIOSH inspection in 1988 and kept pushing

Goodyear to clean up the plant, Polka said. It was still pushing: at the time I first spoke with him, in 2013, Goodyear was resisting union demands to have quarterly pre- and post-shift urine testing for plant workers. Such testing yielded data that could help the company identify the jobs with the highest exposures and adjust processes accordingly. But it was being done only sporadically, even though it was the most accurate way to find out how much ortho-toluidine the workers were absorbing.

The union, Polka acknowledged, was far from perfect and was not immune to pettiness and intemperance. But Local 8–277 had stepped up when it was needed, and its members had stuck together during and after the NIOSH investigation. Workers facing a similar health threat at a non-union plant, Polka speculated, would have been out of luck. "If fifty people in Texas died tomorrow because of some disease at some little, rinky-dink plant that Dickie opened up," he said, "nobody would give two shits about the fifty bodies they're gonna bury."

In the room with the three older men, mostly listening, was Ricky Palmer, a millwright and father of three who had been at the plant seven years. Palmer said he was grateful for what the union had accomplished a quarter-century before. But he thought about what had happened to Dorothy Kowalski and worried about his own children: "Will they get bladder cancer from me bringing [ortho-toluidine] home on my clothes?" Even though there was now a "clean" side of the plant (a locker room in which workers kept their street clothes and showered before they went home) and a "dirty" side (another locker room where their uniforms, work boots, and gloves were donned and removed), the chemical was everywhere.

CHAPTER 16

RAY AND HARRY GET BAD NEWS

O N THE MORNING OF JUNE 28, 1996, Harry Weist came to work at Goodyear and was approached in the maintenance shop by a nervous friend who'd deduced something bad was about to happen. The friend's suspicion was confirmed later that morning. A member of the union's executive board, he attended a meeting called by corporate and learned that Goodyear planned to shut down Department 145—vinyl—at the end of the year. He told Harry, who joined other workers at an all-staff meeting the same day. Harry felt badly for his friends, assuming many of them would soon be unemployed. The *Buffalo News* reported that 166 of 263 jobs at the plant would be eliminated. According to the newspaper, "The Akron, Ohio-based company said the operation is too small to be competitive and does not fit into its core operations. Goodyear said returns on [its polyvinyl chloride] business have been unsatisfactory and the long-term outlook is poor due to the high cost of technology and keeping up equipment."

Harry had nearly nineteen years at the plant and felt safe. Then Goodyear announced it would be cutting people with up to twenty-five years, and Harry's sense of security crumbled. He managed to hang on until January 1997, having been retained to dismantle and mothball the equipment in Department 145. He was among the last to be laid off, and was unhappy with the severance package he received, which, beyond pay and vacation time owed, included a $1,500 bonus and no extended health coverage. It was the first time he'd felt truly disillusioned with the company.

A bad situation was made worse by plant manager Les Carnahan, who despised Harry and the union. The men had a run-in at a company golf tournament in the summer of 1996. Harry, then hot-tempered, threatened to crush Carnahan's skull with a golf club after Carnahan made an ill-advised joke about the impending layoffs. Months later, Harry, newly unemployed, came to the gatehouse to pay his health insurance premium and ran into Carnahan, who said within earshot of others, "Don't get discouraged. You'll be coming back." Carnahan then summoned Harry close and said in a low voice, "You're never coming back, you motherfucker." But Carnahan, too, would lose his job.

Harry was recalled to Goodyear for a few weeks in the spring of 1997 to help install a reactor in Department 245, then was out for almost a year. He found it hard to get decent-paying work, and his income fell by two-thirds. Diane went to work for a company that made jewelry, lamps, and desk sets. Harry signed on with an industrial maintenance contractor that bid on the "hardest, dirtiest jobs." He especially loathed the Cascades paper mill in Niagara Falls, a filthy maw with a sulfurous stink. He worked eighteen or nineteen hours a day in conditions far more treacherous than what he'd known at Goodyear; one of his coworkers was cut in two. Harry felt aimless during his time away from Goodyear. He'd been piling up overtime hours so he could retire at fifty-five and spend more time with Diane and the kids. But he was only forty-one, and now everything was threatened. The layoffs in vinyl had shredded the marriages of some of Harry's friends, who couldn't handle the financial strain, and he wondered if he and Diane would meet the same fate.

Ray and Dottie Kline were supportive during this difficult period. They gave Harry a credit card with no balance and a $16,000 limit and told him he could settle any debt once he was back on his feet. He and Diane charged, and in time repaid, about $12,000 for their mortgage, health insurance, and food. Harry spent his forty-second birthday, March 12, 1998, working at a greenhouse with his father. His mood, already sour, darkened when he got home that afternoon and found a line of cars parked outside. He'd told Diane not to do anything special and now had to play host. When he walked into the house, ready to have it out with Diane, a group of friends shouted, "Surprise!" Diane handed him a certified letter from Goodyear. Harry had been called back to work.

Ray Kline is short, stocky, and guarded, with an uber-dry sense of humor. He speaks in the nasal twang of the central Pennsylvania coal country from which he came. He is disinclined to talk about himself. Harry, by comparison, laughs easily and shares everything. He's known by those close to him as Bud. His catchphrase, spoken in a flat, western New York accent, is "my friend," as in, "You know, my friend, . . ." He's a half-foot taller than Ray. Early in January 1997, the month Harry was laid off, Ray had given a urine specimen as part of a Goodyear screening program. The specimen contained suspicious cells, and Ray underwent a cystoscopy on February 5. He learned the following day that he had bladder cancer; his was the twenty-first case recorded at the plant. Ray handled the news with typical stoicism, though he later admitted in a deposition that he was "mad, despondent [and] scared." Fifty-eight years old at the time of his diagnosis, Ray already had flirted with death, having suffered a heart attack after being struck in the chest with a piece of heavy equipment in 1986. Now he faced six weeks of chemo, which required the insertion of a tube into his ureter, the injection of strong cytotoxic drugs, and the holding of those drugs in his bladder for two hours. The pain, he would say later, nearly brought him to his knees.

Ray, the ultimate company man, tried to tough it out. Harry would hear him vomit in a plant restroom before handing out the day's assignments in the maintenance shop. It was the price Ray paid for working, mostly unprotected, around ortho-toluidine. The liquid had gotten on him when he was repairing pumps and tanks, removing and installing pipes, connecting and disconnecting tank cars. He soaked in it while doing monthly maintenance inside sludge-filled Nailax reactors, a job that took from four to sixteen hours, depending on what needed to be done. Even after he was promoted to maintenance supervisor in 1982, he spent most of his time in Department 245, where he continued to be exposed.

Ray had weathered previous health problems with little complaint. After a heart bypass in 1989, he'd gone hunting alone, shot a buck, and hauled the carcass out of the woods with difficulty, drawing rebukes from his family. He'd had colon surgery for diverticulitis in 1994 and was hospitalized with a viral infection in 1996. In all cases he bounced back. But the cancer had touched him in a way the other ailments hadn't. He was anxious about it most of the time, wondering if it would ever truly be excised or, rather, would go dormant, only to reappear. The cancer did, in fact, come back, necessitating another round of painful, demoralizing treatments.

Having survived the long layoff and returned to work at Goodyear in March of 1998, Harry felt his plan to retire early was back on track. Then came more setbacks. On March 20, 2003, eight days after Harry turned forty-seven, Goodyear floated the idea of selling its chemical division, which included the Niagara Falls plant. The seventy-five workers there were under a "cloud of uncertainty," the *Buffalo News* reported. Goodyear's corporate communications director didn't help matters when he told the newspaper, "We don't have any plans to close the plant, but I can't speculate on what a buyer would do." It turned out to be a false alarm. Niagara Falls and the three other plants in the division, all in Texas, ramped up production and proved their worth to Goodyear, which decided not to sell. Harry had dodged another bullet.

A bigger blow came eighteen months later, on October 20, 2004. Harry had been having trouble urinating and underwent a cystoscopy, which uncovered no evidence of cancer. The urologist performed a biopsy as a precaution and found a tumor, making Harry case No. 37 from the plant. "It was like a brick to the head," Harry recalled, "because I'd watched Ray go through all that cancer and watched my buddies get it." Diane was terrified, fearing she'd lose her husband and her father. It was a low-grade tumor, removed by the doctor, so Harry didn't need chemo. This was a relief because he'd heard Ray talk of "peeing razor blades" after his treatments. Still, there would be physical—and, especially, mental—effects. Harry had trouble getting erections, and Diane learned not to discuss the matter because it would make him cry. Harry developed extreme anxiety about the tumor; it was on his mind constantly. It interfered with his work. It caused him to tune out what Diane and the kids said to him. He grew deathly afraid of cystoscopies, insisting that he be put under general anesthesia each time he had one. He was certain he was going to die. He dreamed about it and would wake up drenched in sweat.

Ray had sued DuPont with great reluctance seven years earlier, fearing it would reflect badly on Goodyear. Harry had no such qualms. Both men chose Steve Wodka to represent them; there had been no reason to look elsewhere. Although Wodka could be demanding, he genuinely cared about his Goodyear clients. Some old-timers said he knew the intricacies of the Niagara Falls plant better than they did.

The Sullivan case had been Wodka's initiation to the peculiar and merciless world of toxic-tort litigation, a world in which anything but the defendant's product caused the plaintiff's illness or death. Smoking, drinking, genetics, atrocious luck—but *not* the product over which you're suing. Wodka tangled over the years with DuPont's lawyer, Paul Jones, and Goodyear's lawyer, Diane Bosse. He found both to be difficult; depositions in which they were involved were constantly interrupted by objections and arguments, which often left deponents annoyed and confused. Wodka was especially puzzled by Bosse's combativeness, given that Goodyear's workers' compensation liabilities for the Niagara Falls bladder-cancer cases could be reduced or eliminated under New York law if DuPont settled the personal-injury claims Wodka had filed. It must have been her nature, Wodka reasoned. Strategically, it made no sense.

Between 1990 and 2005 Wodka's DuPont cases were settled at a regular clip. For this he credited DuPont's in-house counsel, John Bowman, a pragmatist who was receptive to resolving cases in which exposure to ortho-toluidine had occurred prior to 1977, the year DuPont wrote its warning letter to Goodyear. No need to drag things out or hire high-dollar experts to contest a claim DuPont was unlikely to win. Starting with the Sullivan case, Bowman would solicit a settlement demand from Wodka once the plaintiff's medical and employment records had been collected and the plaintiff had been deposed by DuPont. Things would move relatively quickly from that point. Ray Kline's case, for example, was filed on April 18, 1997, and was closed out by July 26, 1999.

It was Bowman who'd written a memo urging DuPont executives to do something about perfluorooctanoic acid, or PFOA, an environmentally persistent chemical used in non-stick cookware that contaminated the Ohio River near the company's Washington Works manufacturing complex in West Virginia. Part of a family of chemicals known as per- and polyfluoroalkyl substances, or PFAS, PFOA has been linked to cancer and other health effects and is present in the blood of almost every American. "We are going to spend millions to defend these lawsuits and have the additional threat of punitive damages hanging over our head," Bowman wrote in 2000. "Getting out in front and acting responsibly can undercut and reduce the potential for punitives." His advice went unheeded. After Bowman retired in 2005, Wodka went seven years without settling a DuPont case. The company's new counsel, Janet Bivins, insisted on fighting every lawsuit and spent lavishly on experts and outside lawyers. Every point was contested: when Wodka would propose what he thought to be a reasonable discovery schedule of six to nine months, DuPont would want years. Among the thirty bladder-cancer claims Wodka brought against the company—twenty-eight from Goodyear in Niagara Falls, two from other plants—Joe Nicastro's came closest to going to trial. DuPont filed nine motions to dismiss all or part of that case, or to challenge the proposed testimony of the plaintiff's experts. In the two months prior to the scheduled trial date, DuPont filed twenty-nine motions to bar certain types of evidence.

Like all personal-injury lawyers alleging disease from a chemical exposure, Wodka was forced to deal with the "anything but my product" defense. The Nicastro depositions illustrate this phenomenon vividly. During the three sessions, lawyers for DuPont and the other defendants zeroed in on Joe's habits and family history, sidestepping the fact that he'd worked for decades with at least two virulent bladder carcinogens, ortho-toluidine and beta-naphthylamine. Joe testified that he started smoking when he was sixteen and quit when he was sixty-four. He'd stop for "weeks, months," and then pick up the habit again. What brands had he smoked? the lawyers wanted to know. Filtered Parliaments, Marlboros, and Newports. How many cigarettes a day? Ten to fifteen, Joe said. But wait, one lawyer said. Joe had once told a doctor he'd gone through a pack a day—twenty cigarettes. "I might have," Joe admitted. This dispute—had he smoked twenty cigarettes or fewer?—went on for several minutes. Hadn't Joe's first wife and mother been smokers? They had. This, one of the lawyers pointed out, meant that Joe had been exposed to secondhand smoke over a period of years. Joe replied that both his hematologist and his oncologist believed his illness had been spawned by the chemicals he'd worked with in New Jersey.

There were also questions about Joe's beer consumption, as if a few cans of Budweiser a day could somehow compete with skin-penetrating ortho-toluidine as a mortal threat to the bladder. Joe wearily acknowledged that he'd reaped the benefits of the reward system at Patent Chemical as a young man. When tank trucks loaded with dye were shipped out, the plant owner would buy several cases of beer for the workers. Joe drank mainly in the summer, when his colleagues would sneak in beer and sell it, "and if you were thirsty, you bought a can of beer."

"You did drink on the job at Morton, did you not?" a lawyer asked.

"Yes," Joe replied.

"In fact, while on the job at Morton, while exposed to all the chemicals that you're exposed to, you could drink as much beer as you wanted, is that right?"

"Yes."

And what about his brother, John, who, like Joe, had worked at their father's trucking company and, like Joe, had developed bladder cancer?

"Did your brother John ever work at Pfister?"

"No."

"Did he ever work at Patent?"

"No."

"Did he ever work at Morton?"

"No."

"Did he ever work at Rohm and Haas?"

"No."

Hadn't Joe's mother and father died of cancer? They had.

"And none of them worked at Pfister or Patent or Morton or Rohm and Haas, correct?"

"Correct, and they all got cancer. Isn't that amazing?"

Oh, and why hadn't Joe submitted his retiree urine samples to Rohm and Haas in 2005, 2006, and 2007? Joe said he'd received the test kits, put them in his garage, and forgotten about them. The implication was that it was partly Joe's fault his cancer hadn't been discovered earlier, when it might have been more treatable. The counterargument was that the suppliers—especially DuPont—might have been a bit more forthcoming about the carcinogenicity of the chemicals they sold to Joe's employers all those years. Had that happened, exposures might have been curtailed sooner than they were.

Pam Nicastro was present for all three of her husband's depositions and fumed as the defense lawyers asked Joe what she perceived to be moronic, insulting questions. She was deposed herself on January 16, 2009. In the videotaped session she acknowledges marrying Joe on August 25, 1994, and says in response to a question that the couple has experienced no "marital problems" (as if such problems would have had anything to do with Joe's cancer). Had she ever smoked? Yes, she'd started at sixteen but quit before she had her first child in 1980.

How had she felt on the day Joe was diagnosed with bladder cancer? "It felt like someone pulled the rug from underneath me," she says. "I couldn't breathe." And what sort of care did her husband require? This question seems to irritate her, and she starts to cry. "Do you want to know what I do for him every day?" she asks the lawyer sharply. "Do you want to know the gory details?" The lawyer drops the line of questioning and the deposition ends soon after.

Harry Weist had experienced similar gamesmanship in his own case against DuPont and had to summon every ounce of his self-control to keep from losing his temper. DuPont's medical experts played down his condition, insisting his tumor was so negligible that there was almost no chance it would recur or progress. The experts were wrong. After lurking for fifteen years, Harry's tumor would re-emerge in 2019 as high-grade, putting him at risk of muscle invasion and, therefore, death.

At the urging of a few friends, Harry ran for and was elected vice president and recording secretary of the local in 2004. Once he took office, he became the de facto counselor for Goodyear retirees diagnosed with bladder cancer, a role he didn't welcome given his own fragile state.

He heard from one former colleague whose bladder had to be removed and who died a terrible death. He heard from another—a tough guy who intimidated the bosses—who also died. He heard from a stricken former supervisor. He briefed the men on his lawsuit and gave them Steve Wodka's contact information. He concluded that Goodyear didn't care about any of them.

As Harry was battling his inner demons and DuPont's lawyers, more turmoil unfolded within the Weist family. Harry's seventy-eight-year-old father, Daniel, died of a type of brain cancer known as glioblastoma multiforme on March 8, 2007, and was buried on March 12, Harry's fifty-first birthday. The two of them had been exceedingly close, sharing a passion for gardening and bowling, and Harry was heartsick.

This period in Harry's life was not joyless. He and Diane had first visited Las Vegas in 1993, winning about $4,000 between them on the slot machines and craps tables, and became so smitten with the city that they went back every year. They celebrated each of their children's twenty-first birthdays there and spent freely, once dropping $3,500 in a single, boozy evening at the Mirage. But Harry's cancer, and his lawsuit, were always present, even if they occasionally were pushed to the background. Harry kept expecting the case to be settled and kept having his hopes dashed. One February during Super Bowl week in 2008, Wodka called him in Las Vegas at 6 a.m. local time—Harry had yet to go to bed—to say a settlement seemed imminent, but then negotiations fell through. Wodka was livid.

In May of 2011, Harry retired from Goodyear after more than thirty-three years with the company. He'd gotten out, as planned, at fifty-five. But, having worked eighty-hour weeks, he found it hard to relax at home. Diane and the kids worried about him, told him he looked drained. The lawsuit weighed on him. In his first summer away from Goodyear, Harry, never adept at idling, found an outlet for his restlessness and natural gregariousness in a job that paid $8.50 an hour. He delivered auto parts to garages in the Niagara Falls area and was popular with the customers, one of whom asked him to record a radio testimonial. Finally, in November of 2012, his lawsuit against DuPont was settled. It had taken eight years. The settlement, the terms of which are confidential, inspired the Weists to start looking for a house in Las Vegas, where they planned to spend each winter and spring. In 2013 they bought a tri-level with a pool on South Mojave Road, ten minutes from the Strip. Harry's cancer appeared to be in check, the lawsuit was behind them, and the kids—Holly and Harry Jr. in Colorado Springs, Kristan in Las Vegas—were a quick car ride or flight away. They held onto the farm in New York at Diane's insistence, though it was a financial sinkhole. Harry was finally able to relax, in a relative sense. He spent his time doing yard work and maintaining the pool. He and Diane went to the casinos and ate out. "Damn," Harry recalled. "Life was good."

CHAPTER 17

WODKA FORTIFIES HIS CASE AGAINST DUPONT

SEVEN MONTHS BEFORE Harry Weist's lawsuit against DuPont was settled, Steve Wodka took deposition testimony in another case that
laid bare the company's duplicity. The witness was James Medaris,
then eighty-one, who had been a supervisor in the hydrogen reduction
building at Chambers Works. It was in this building that ortho-toluidine
and other aromatic amines were made. Testifying in the Joe Nicastro
case, Medaris acknowledged that the amines, known for their ability
to cause cyanosis, the skin-bluing oxygen deficiency, were handled with
extreme caution. Workers who loaded the liquids into 55-gallon drums
wore impermeable butyl-rubber suits, gloves, and boots and breathed
supplied air, Medaris said. They used glove boxes to prevent even the
most trifling exposures to the skin. "All persons handling the product
must wear protective clothing designed to avoid skin contact, and efficient ventilation must be provided in all work areas," company policy
stated. "Each person leaving the work area is required to wash his face,
neck, hands and forearms." These measures were in place when Medaris
started in the hydrogen reduction building in 1965, he said, and were
still in place when he retired thirty years later. For the first twenty-one
years of this period, Goodyear workers in street clothes were getting
splashed and occasionally soaked with ortho-toluidine. Goodyear didn't
start providing uniforms and laundering them until 1986. Had DuPont

communicated to Goodyear just how treacherous the chemical was, how delicately it was being handled at Chambers Works? It hadn't.

Wodka struck evidentiary gold again in 2016, when he deposed Barbara Dawson, who had been the industrial hygienist at Chambers Works for a decade before becoming DuPont's "global occupational health competency leader." Dawson confirmed that since the 1940s, ortho-toluidine had been considered a "no-contact chemical" at DuPont because of its cyanotic effects. She acknowledged that by 1986, DuPont allowed no amount of the chemical to be present in a worker's urine. If it was, the worker was to be pulled off the factory floor and sent to the medical office for questioning.

Still, the company insisted in a 1986 memo, there were "no scientific literature reports or Chambers Works experience that would indicate that [ortho-toluidine] causes cancer in humans." Wodka pointed out that there had been hundreds of cases of bladder cancer at the plant. "Yes," Dawson replied. "They were attributed to other chemicals, however." She was referring to beta-naphthylamine and benzidine.

In 1990, Dawson wrote disingenuously that OSHA's 5-parts-per-million air exposure limit for ortho-toluidine was considered "protective of our employees' health and will be continued." The news of mounting bladder-cancer cases out of Goodyear in Niagara Falls, in the form of a 1992 NIOSH report, shook DuPont from its slumber. The following year, Thomas Nelson, an industrial hygienist at corporate headquarters, made a calculation proving that the legal exposure limit for ortho-toluidine provided workers little protection. Someone exposed to 5 ppm of the chemical over eight hours, Nelson estimated, would have a urine level of 20 milligrams per liter—eight times what DuPont had decided was acceptable in 1953 and thirty-seven times the highest level NIOSH had detected in a Goodyear worker. DuPont never shared this finding with Goodyear or federal regulators.

And yet DuPont continued to maintain that it had been a responsible corporate citizen. On a dank August morning in 2018, Wodka showed up on the seventh floor of the Robert H. Jackson United States Courthouse in Buffalo to argue once again over culpability for the Goodyear disaster. DuPont's outside counsel, the slight, bespectacled

Alan Wishnoff, stood as US Magistrate Judge H. Kenneth Schroeder peered down from the bench. DuPont had filed a motion to dismiss one of Wodka's cases—that of Goodyear worker Douglas Moss and his wife, Suzanne. Schroeder had to decide whether to recommend to a US district judge that the motion be granted. The Mosses sat in the first row of the courtroom's sparsely occupied spectator section. Wodka took notes at a table to Wishnoff's left.

Wishnoff argued that *Moss v. DuPont* should be thrown out on summary judgment, and he seemed to score points early on. Two years before Doug Moss started at the Goodyear plant in 1979, DuPont notified all its customers by letter that ortho-toluidine was a "suspected carcinogen" and that workers likely to encounter it should be sealed in rubber suits and breathe supplied air, Wishnoff said. Goodyear, a "large, sophisticated company" that bought millions of pounds of the chemical from DuPont and other suppliers, should have acted on the information but didn't. "We're not talking about a mom-and-pop grocery store here," Wishnoff said. Schroeder appeared to appreciate the point. Then Wodka stood and began his rebuttal. The 1977 letter that DuPont was so proud of was actually a model of equivocation, he said, a warning so feeble it was no wonder Goodyear didn't take it seriously. Yet, by the 1980s, DuPont itself was so alarmed by ortho-toluidine's ability to penetrate the skin and target the bladder that it adopted a "zero-tolerance" policy at Chambers Works: even a trace detected in a worker's urine triggered an in-house investigation. "They never told Goodyear about that," Wodka said, his voice rising. Another of his Goodyear clients, James Sarkees, sat by himself in the last row of the spectator section during the arguments. Sarkees had worked at the Niagara Falls plant for seven months in 1974; now, at sixty-four, he had bladder cancer.

The following July, I attended a hearing in Buffalo's dingy Ellicott Square Building on a workers' compensation claim Wodka had brought against Goodyear on behalf of Guy Mort, who was diagnosed with bladder cancer in 2015 and still worked at the Niagara Falls plant. Unlike the men who had sued DuPont, Mort was seeking lifetime medical care, nothing more. Nonetheless, he met resistance. Goodyear's position was that no one who started working at the plant after 1995, as did Mort, could have gotten bladder cancer from ortho-toluidine because exposures to the chemical were so tightly controlled from that point on. This seemed reasonable until one looked at the data. In October 2018, results from the latest round of urine testing provided by the company came in. Despite the controls installed by Goodyear, the average post-shift amount of ortho-toluidine in a Niagara Falls worker's urine was still forty-two times the background level—what anyone not occupationally exposed to the chemical would be expected to have. This suggested that the cancer threat had not disappeared.

Mort—stocky and bald—seemed skittish before the hearing and did not want to be interviewed afterward. With him was Joe White, then president of United Steelworkers Local 4-277, which represents the Goodyear workers in Niagara Falls. White, too, seemed wary. Wodka later explained the union leader's dilemma: on one hand, it was his duty to advocate for safe working conditions; on the other hand, White feared that if he pushed too hard, the company would close the plant, already operating with a bare-bones crew. Wodka opened the three-hour hearing by stating that Mort had developed bladder cancer "as a result of his occupational exposure to a chemical called ortho-toluidine. Because of this malignancy, Mr. Mort will remain under the care of a urologist for the rest of his life." Mort had started at the Goodyear plant in 2004, worked in production and maintenance, and was diagnosed with the disease through a company urine-screening program eleven years later. The cause of Mort's ailment was no mystery, Wodka said, noting that ortho-toluidine, easily absorbed through the skin, is classified as a "known human carcinogen" by the premier American and international health authorities. Tests showed levels of the chemical in Mort's urine were more than thirty times higher than what would be found in the general population.

Goodyear had been one of the biggest users of ortho-toluidine in the United States for decades and was still using between five and eight million pounds of it per year as of 2017, Wodka said. He showed Mort photographs of stained, hulking vessels and asked if he could identify them. Yes, Mort said—the tanks were known as Premix 1 and Premix 2, where ortho-toluidine and other chemicals were held before being sent to the reactors in which Nailax was made. The dark streaks on the tanks were remnants of spills, Mort said; in fact, "just six months ago we had one of these overflow." Mort also confirmed that he had worn ultrathin, nitrile gloves when taking chemical samples—gloves not recommended for handling ortho-toluidine, even intermittently, for more than fourteen minutes. Had he known about this time limit? "No," Mort said. Had he worn them for more than fourteen minutes? He had.

Wodka had made his point: employee exposures at Goodyear were still occurring. It was agreed that Goodyear would present its rebuttal at a hearing on August 30. A few weeks before that date, the company dropped its challenge to Mort's claim. The decision issued by the state Workers' Compensation Board in September found that Mort "has an occupational disease involving bladder cancer as a result of chemical exposure." Goodyear and its insurance carrier objected to the phrase "as a result of chemical exposure" and asked that an amended decision be issued without it. It was; Wodka didn't protest because he'd gotten what his client wanted.

CHAPTER 18

CHEMICALS ARE OUT
OF CONTROL

A S FEARSOME AS THE CANCER OUTBREAK at Goodyear has been, it's hardly the only workplace to be compromised by chemicals. Because of union and government intervention, we simply know more about it than most. Two months after the Democrats took over the House of Representatives in 2019, the new chairman of that body's Committee on Energy and Commerce, Frank Pallone Jr. of New Jersey, presided over a hearing on the Environmental Protection Agency's inept management of chemical risks to workers. "Clearly, our track record of protecting workers is appalling," Pallone said in his opening statement. He gave as an example methylene chloride, a paint stripper the Obama administration had tried to ban because it kept killing people, mostly workers in confined spaces. The Trump EPA had decided workers should still be able to use it (though consumers would be prohibited from doing so later that year). Pallone noted that a worker somewhere on the planet was struck down by chemicals every fifteen seconds. "To put that in perspective," the congressman said, "by the time my five minutes are up, toxic exposures will have killed 20 workers worldwide." Adam Finkel, a former OSHA health standards director who advised Steve Wodka in the DuPont litigation, testified that his analysis of some three million OSHA air samples showed occupational levels of certain chemicals were up to one million times higher than what would be found in the ambient environment. It's reasonable to assume workers in a factory know "they face more risk than the general population," Finkel testified, "but a million times more?" While exposures to asbestos, benzene, and other old poisons had fallen, he said, "the overall problem is not decreasing significantly, as other exposures stay high and new substances replace older ones, sometimes with equal or greater toxicity."

Witness Jeaneen McGinnis, a retired Chrysler assembly worker from Huntsville, Alabama, reinforced Finkel's point. She'd started at the plant in 1983, "overjoyed" to find work in a place where jobs were scarce. The plant made radios, air bags, odometers, and other automotive components. McGinnis assembled circuitry for dashboard instruments including the speedometer, fuel gauge, and check-engine lights, standing without protective gear as each circuit board was "passed across a wave of molten solder." Workers on the line inhaled the acrid fumes and absorbed substances that had not been identified to them, McGinnis said. They didn't even wear gloves. Conditions improved after she transferred to a newly opened Chrysler plant in Madison, Alabama, in the early 1990s, but even there, workers endured "inadequate ventilation, insufficient training on how to handle the chemicals and unfamiliarity with the chemicals we were using." The solvent used to clean the resin off circuit boards was trichloroethylene—TCE—a known human carcinogen. McGinnis moved off the line in 2003 to become a benefit representative for the United Auto Workers. She attended a weekly luncheon with 90 to 130 retirees from the Huntsville plant. "I am accustomed to showing up and not seeing the person that I sat next to from the previous lunch because they are no longer with us," McGinnis testified.

The takeaway from the hearing was that even in the twenty-first century, hardly the dawn of the industrial age, many blue-collar workers were still being treated like vassals, easily replaced if they got sick or made trouble. Wilhelm Hueper, the DuPont gadfly, had predicted what was to come many years before, in his 1942 textbook, *Occupational Tumors and Allied Diseases*. "Although today a great deal of evidence on the causes of occupational cancer and the conditions under which it develops is available, relatively little effort has been made to spread this knowledge among parties who are vitally concerned with this information, namely the managers and physicians of industrial concerns, although they carry the main load of controlling this preventable type of cancer," he wrote. "This situation is the more deplorable because the present rapid development of industry tends to extend the hazards of the known industrial carcinogenic agents while at the same time creating new ones which may cause the appearance of cancers ten to twenty years hence, unless concerted efforts are made in time to discover and control such dangers by appropriate countermeasures."

Bladder cancer in the aniline dye industry was a perfect example of this phenomenon. Although the disease had appeared in German workers in 1895, the suspected agents, beta-naphthylamine and benzidine, were handled with no special care at DuPont's Chambers Works in New Jersey several decades later. In 1932, fifteen years after the American dye industry had reached "large proportions," the first bladder tumors appeared in the plant's workers. "Within the short period of six years thereafter," Hueper wrote, "almost 100 cases of this industrial cancer occurring in men of relative youth were placed on record from one single chemical concern"—DuPont—"attesting to the presence of a relatively massive exposure of the workers to the causative agents." By comparison, Hueper noted, only about three hundred occupational bladder cases were recorded in the entire German chemical industry over forty years, from 1895 to 1935. It's no wonder he was persona non grata at DuPont.

New threats had already been identified by American investigators, Hueper wrote. Carbon tetrachloride, a common solvent, and ethyl carbonate, used in anesthesia, had been linked to liver and lung cancer, respectively, in mice. Rats fed a solvent called diethylene glycol had developed bladder tumors. "The total number of occupational cancers placed on record is small when compared with the large number of cancers of unknown etiology," Hueper observed—again, in 1942. "It is obvious, however, that the latter group contains an undetermined number of cancers of occupational origin, as many of them either are not properly recognized or are not made a matter of public information."

Another prophet emerged in 1946: toxicologist Henry F. Smyth Jr., who directed an industrial hygiene fellowship program for the Union Carbide Corporation at what was then the Mellon Institute in Pittsburgh.

"It is clearly the duty of a manufacturer to delay production of a chemical until the health hazards are well enough defined so that protection of his workmen is possible," Smyth wrote in the *West Virginia Medical Journal.* "It is also his duty not to sell a chemical for an application in which it would endanger the health of the public, and to inform customers, by proper labeling and otherwise, of the hazards of the compounds they buy."

Hueper's and Smyth's warnings came at the start of a period of explosive growth in chemical production. The United States had begun exploiting its ample supplies of crude oil and natural gas during the war to produce petrochemicals such as synthetic rubber. "World War II resulted in the physical destruction of a significant portion of the German chemical industry," Ralph Landau and Ashish Arora wrote in a 1999 article for the journal *Business Economics.* "The U.S. industry was now using petrochemicals to produce fibers, plastics, and many other products, while dyestuffs shrank in importance." By 1950, Landau and Arora wrote, almost half of the nation's organic chemicals had roots in fossil fuels; by 1960 it was nearly 90 percent.

The companies that profited from this transition knew there was a human price; they just didn't publicize it. In 1948 came the now-infamous (in trial-lawyer circles) toxicological review of benzene, prepared for the American Petroleum Institute (API) by the Harvard School of Public Health's Philip Drinker, which challenged the safety of the 100-parts-per-million exposure limit then in effect. "A limit of 50 ppm is strongly recommended, particularly where exposures are recurrent," Drinker advised. However, "inasmuch as the body develops no tolerance to benzene, and as there is a wide variation in individual susceptibility, it is generally considered that the only absolutely safe concentration for benzene is zero."

In 1950, C. H. Hine, a physician and consulting toxicologist, wrote a confidential memo for Shell Development Company that turned up years later in litigation discovery. Its title was "Certain Problems of Environmental Cancer in the Petroleum Industry." While he saw "no cause for immediate concern," Hine wrote, "there is little doubt that there is a quantitative widening of the spectrum of environmental carcinogens."

Many were caused by "exposure to ill-defined mixtures of organic chemicals, the carcinogenic components of which are incompletely known." The API had begun a survey of members that kept good medical records, seeking data on cancer in the workforce. "Information of a highly confidential type has reached me that Esso has conducted such a survey and has obtained information which is causing considerable concern," Hine remarked without elaborating. Attached to the memo was a table listing confirmed and suspected carcinogens found in the industry, and the organs they were known or thought to target. Another table listed substances that had elicited pre-cancerous reactions. "In only relatively few instances can the origin of environmental cancer be traced to contact with well-defined chemical agents possessing established carcinogenic qualities," Hine wrote. "Among such compounds are arsenic, benzol (benzene), and aromatic amines, in addition to radio-active elements." It would be another seven years before Goodyear would begin using one of the amines, ortho-toluidine, at its Niagara Falls plant. Had members of Goodyear's medical staff reviewed the literature on this diabolical family of chemicals? Had they read, for example, the 1934 article in the *Journal of Industrial Hygiene* in which Hueper foretold a wave of bladder cancer in plants that used these agents? Had they seen the 1949 article in the *British Journal of Industrial Medicine* in which M. W. Goldblatt concluded "there is such a thing as aromatic amine cancer"? If they had, no one told the workers in Department 245.

In the typical American factory of the 1950s, ignorance of chemical hazards was the norm. For many workers it would prove fatal. An underappreciated figure of this era was Herbert Abrams, a Chicago-born physician who worked for the US Public Health Service and the California Department of Public Health. While at the latter he saw the on-the-ground impacts of pesticide poisoning among farmworkers and proposed the then-radical idea of forcing manufacturers to label their products not with meaningless trade names but with a list of the chemicals used to make those products, and their effects on the human body. Abrams also believed workers studied for ill effects of toxic substances had as much right to see the findings as did their employers. This, too, was highly unusual for the time. The Occupational Safety and Health

Administration did not exist, and the states were the primary regulators of workplaces. Many of the states "had laws or legally binding regulations that forbade government officials from revealing to anyone other than the employer the information gained in the course of inspections and investigations or that barred the admission of their findings into workers' compensation cases or lawsuits," Alan Derickson wrote in a profile of Abrams in the *American Journal of Public Health*. Abrams became a consultant to the International Chemical Workers Union and in 1953 remarked in a column for the union newspaper that "if you are a low income industrial worker, and especially if your skin is not white, your chances for health and long life are not as good as your fellow men in the higher income brackets." After learning of a bladder-cancer cluster at a Monsanto plant in St. Louis that had been concealed from the plant's workers, Abrams "arranged for the National Cancer Institute to set up an exhibit [at a union convention] on occupational cancer that included a large supply of Wilhelm Hueper's booklet surveying the problem," Derickson wrote. Abrams urged workers to approach the health and safety committees of their locals with any health concerns and, if that didn't work, file a complaint requesting a state investigation. He lamented that "many thousands of hazardous chemicals today come under no regulatory laws" and maintained that "unless the worker himself knows what he is doing and handling and how to protect himself, all other measures will fail." His teachings presaged the development of OSHA's Hazard Communication Standard, a process that began in 1974 and came to fruition for manufacturing workers in 1983. Workers in all industries were covered by 1987.

Herbert Abrams flew mostly under the American public's radar. This was not the case with biologist Rachel Carson, whose apocalyptic book on the proliferation of DDT and other pesticides, *Silent Spring*, created an outcry when it was published in 1962. "If we are going to live so intimately with these chemicals, eating and drinking them, taking them into the very marrow of our bones, we had better know something about their nature and their power," Carson wrote. President John F. Kennedy read excerpts of the book in the *New Yorker* and was so shaken that he established a special panel to investigate Carson's dire predictions.

The panel's report in May 1963 validated her findings and bolstered her credibility, which had come under fierce attack by the chemical industry. ("If man were to faithfully follow the teachings of Miss Carson," an executive with American Cyanamid had griped, "we would return to the Dark Ages, and the insects and diseases and vermin would once again inherit the earth." Monsanto had handed out a parody brochure, playing off *Silent Spring*, entitled "The Desolate Year.")

Carson's book roused the budding environmental movement. In 1963, Congress passed the first version of the Clean Air Act. Lyndon Johnson warned in 1965 of an ominous carbon dioxide buildup in the atmosphere; the following year there were Senate hearings on the perils of leaded gasoline. In 1969 came two disasters: a massive oil spill in the Pacific off Santa Barbara, California, and a conflagration on the oil- and chemical-fouled Cuyahoga River in Cleveland. In July 1970, not quite three months after the first Earth Day, Richard Nixon—who had barely mentioned the environment during his 1968 presidential campaign but was attuned to public sentiment about worsening pollution—presented Congress with a plan to consolidate heretofore scattered regulatory functions under a single Environmental Protection Agency. Lawmakers embraced the idea, and the EPA's first administrator, William D. Ruckelshaus, was sworn in on December 4 of that year—twenty-five days before Nixon signed the Occupational Safety and Health Act into law.

In February 1971, Nixon proposed legislation giving the EPA the ability to regulate chemicals, not as components of air or water pollution but as stand-alone threats to public health. He used as the basis for his Toxic Substances Control Act (TSCA) a sobering report from the Council on Environmental Quality that would be made public two months later. The council found that two million chemical compounds existed and thousands more were being discovered each year. "Most new compounds are laboratory curiosities that will never be produced commercially," it reported. "However, several hundred of these new chemicals are introduced into commercial use annually." Testing had "largely been confined to their acute effects, and knowledge of the chronic, long-term effects, such as genetic mutation, is inadequate." There was a need for new legal authority, the council concluded. "Our awareness of environmental threats, our ability to screen and test sub-stances for adverse effects, and our capability to monitor and predict, although inadequate, are sufficiently developed that we need no longer remain in a purely reactive posture with respect to toxic substances. We should no longer be limited to repairing the damage after it has been done; nor should we continue to allow the entire population or the entire environment to be used as a laboratory."

J. Clarence Davies, a member of the council's staff who had authored a provocative book called *The Politics of Pollution* while on the faculty at Princeton, explained the council's thinking in an oral history interview for the Science History Institute in 2009: "I recommended . . . that there should be some kind of institutionalized way, presumably through a law, to deal with new chemicals. Because clearly, there was a constant parade of chemical crises, but each one was treated as if it was a unique event. There was no cumulative learning at all, and no institutionalization of how to deal with these crises. I mean, it was the 'chemical of the month' syndrome. . . . Every month there was something new, and it was treated as if it had no relationship to other things that had gone before."

Nixon "hated environmentalists," Davies said, but insisted, "We've got to do something about the environment." The toxics bill was circu-lated among government agencies before being sent to Capitol Hill, and was poorly received by the Department of Commerce, which, Davies recalled, said it would be "quite happy doing without a chemicals bill al-together, and what was a conservative Republican administration doing proposing something like this? They did their best to kill it outright but didn't succeed." In negotiations with Davies, however, the Commerce Department's general counsel, James T. Lynn, managed to get the bill watered down so it would be more palatable to industry. "I think most of the stuff that, in effect, made it clear that the burden of proof was on the EPA and not the manufacturer was his doing," Davies said.

TSCA failed to find political backing and didn't pass until 1976. By then it was a shambles, full of loopholes that grandfathered in some sixty-two thousand chemicals already on the market. Companies were to voluntarily report to the EPA any adverse health or environmental effects they had seen. The agency, suffice it to say, was not inundated with such information. The year after the law took effect, biologist Barry Commoner wrote a lengthy piece for the *New York Times* with the headline "The Promise and Perils of Petrochemicals." In it, Commoner offered a dystopian view of what the industry had wrought:

> The petrochemical industry's products, made chiefly out of crude oil and natural gas, make up a marvelous catalogue of useful materials: cloth with the sheen of silk or the fuzziness of wool; cables stronger than steel; synthetics with the elasticity of rubber, the flexibility of leather, the lightness of paper, or the workableness of wood; detergents that wash as well as soap without curdling in hard water; chemicals that can kill dandelions, but not grass; repel mosquitoes, but not people; diminish sniffles, reduce blood pressure, or cure tuberculosis.
>
> But something has gone wrong. Increasingly, the chemist succeeds, brilliantly, in synthesizing a new, useful, highly competitive substance, only to have it cast aside because of its biological hazards: Food dyes and fire-retardants for children's sleepwear are banned because they may cause cancer; a new industry to produce plastic soda bottles, developed at a cost of $50 million, comes to an abrupt halt as the Food and Drug Administration discovers that a chemical which may leach out of the bottles causes tumors in mice; pesticides are taken off the market because they kill fish and wildlife; firemen would like to ban plastic building materials because they produce toxic fumes when they burn.

"The most chilling prospect," Commoner went on, "is that much of the cancer problem in the United States may eventually be laid at the door of the petrochemical industry." He noted that the nation's highest incidence of bladder cancer could be found in Salem County, New Jersey. That was the site of DuPont's Chambers Works, though Commoner didn't mention it by name. TSCA, he wrote, could have one of two outcomes. Its enforcement could become "bogged down in niggling debates over what constitutes 'an unreasonable risk of injury' and over the comparable benefits of each particular substance." Or, it could open a "political Pandora's box, providing an arena in which the public will be able to intervene effectively in a process that is normally the exclusive domain of industrial managers: decisions about what to produce and how to produce." The former has prevailed.

Congress amended TSCA in 2016 to try to shift the burden of proof from the EPA to chemical manufacturers, who would have to show that new products didn't present "unreasonable" risks. Ten of the most menacing existing chemicals, including methylene chloride, would undergo exhaustive risk evaluations, followed by a group of twenty slightly-less-scary substances, then another twenty, and so on. In fact, the EPA found risks it deemed unacceptable for all ten members of that first group; as of early 2022, it had proposed a ban on imports of asbestos, used exclusively by the chlorine industry, and was considering regulations to address the others. That's progress, of sorts.

The trouble is, 86,631 chemicals were in the EPA's TSCA inventory as of that time. Of these, 42,039 were considered "active." Few will ever be tested for safety. The late Sheldon Krimsky, an environmental ethicist at Tufts University, framed the problem this way in 2017: "The law requires the EPA to have 10 ongoing risk evaluations in the first 180 days and 20 within 3.5 years. Let us assume it will have to undertake risk evaluations for 10% of the existing chemicals—that's 8,500 in groups of 20 to be completed every 3.5 years. That would take about 1,500 years to complete. That is not a very encouraging outcome. . . . With a priority list of 500 chemicals a year and a 3-year completion time, the task could be completed in 50 years."

Fifty years to get through what presumably are the five hundred most widely produced, threatening chemicals? We're the victims of our own misguided system, Krimsky argued. "It may help to remind ourselves that operating under the assumption that a substance is safe until proven otherwise, it has taken 20–25 years after a chemical has entered the commercial market to strictly regulate or prohibit its use, as in the case of lead, PCBs, asbestos, and dichlorodiphenyltrichloroethane (DDT)," he wrote in *PLOS Biology*. "Manufacturers challenge the science and use the uncertainties in the risk assessment opportunistically to demand further studies until there is unimpeachable consensus that the chemical is a public health hazard. If we operated under the assumption that a substance is unsafe until proven otherwise, the onus would be reversed:

the manufacturers would have to spend decades in research to remove all uncertainty and demonstrate that a chemical was unimpeachably safe—a true precautionary approach. Society would not be spending a millennium playing catch-up with the unknown risks from having adopted an approach that favors commerce over health."

The Trump EPA undermined the review process by making assumptions that made substances appear safer than they probably were. One assumption, for example, was that workers would always wear personal protective equipment, or PPE, when handling poisons. That, experts knew, didn't happen in the real world. The EPA also argued that chemical exposures on the job were already policed by OSHA; therefore, another cop wasn't needed. Another fallacy. The Biden EPA has tried to rectify the sins of the Trump era, and appears to have the chemical program back on course. When it assesses a chemical's risk, for example, the agency now looks at all relevant exposure pathways—air, water, waste. This is especially important for people who live near oil refineries, chemical plants, and other dirty industrial operations. The Biden EPA also jettisoned the hypothesis that workers always wear PPE.

Still, industry stands prepared to challenge any finding it doesn't like. For example, the American Chemistry Council (ACC), the chemical industry's main trade group, complained in 2022 about an EPA draft risk assessment of formaldehyde that linked the chemical, used in building materials and household products, to at least three cancers. Although formaldehyde was already considered a human carcinogen by the International Agency for Research on Cancer and the US National Toxicology Program, the ACC, which spent $14.2 million on lobbying that year, complained in a press release that the EPA document could be used "to guide regulations or to set policy at any level of government." The implication was that the government should keep studying this known cancer producer until . . . well, when, exactly? Some environmental advocates believe the EPA is still giving such groups, and their lawyers, too much sway, still cloaking chemical reviews in needless secrecy, and still bending to industry demands for quick decisions. All of this could hurt workers as well as the general public; if history is a guide, it will probably hurt workers more.

CHAPTER 19

OLD SCOURGES REVISITED

F OR AMERICAN WORKERS, the health scourges of the past can seem uncomfortably close at hand. That's due largely to our collective amnesia about the toxic dusts, fumes, vapors, and liquids that exterminated too many of our ancestors and our flawed assumption that *surely someone has taken care of that by now.* This is often not the case. The United States, for example, remains caught in what scientists and physicians at a New York conference in 1990 predicted would be a third wave of asbestos-related disease. The first two waves were bad enough: the fire-resistant mineral had killed asbestos miners, millers, and manufacturing workers, then had taken insulators and shipbuilders. Eventually, it would be roused from its dormant state in pipes, ceiling tiles, and automobile brakes and kill again, even at very low exposures. The killer in most instances would be mesothelioma, which barely registered in the United States and Europe in the early twentieth century and wasn't diagnosed at Massachusetts General Hospital, a bellwether for industrial disease, until 1946. Irving Selikoff, two years away from his own death from pancreatic cancer, convened the New York conference, which came on the heels of papers in the *New England Journal of Medicine* and *Science* that argued for simply keeping asbestos in place. Authored by scientists with industry ties, the papers mocked Selikoff as an alarmist and insisted that, if left undisturbed, asbestos—especially the most common variety, white, or chrysotile—posed little risk.

Presentations in New York suggested otherwise. Philip Landrigan, a pediatrician who worked with Selikoff at Mount Sinai, cited an EPA risk assessment that predicted a thousand people exposed to asbestos as schoolchildren would die prematurely of mesothelioma or lung cancer over the next thirty years. "The result of our collective failure is reflected in the fact that asbestos is widespread in schools and other buildings today," he said. Its dormancy could not be assured. David Lilienfeld, also of Mount Sinai, offered case reports on four teachers—two men, two women—who had developed mesothelioma merely by showing up for work. Their ages were 43, 52, 60, and 64. All but the 64-year-old had died. A team from the Wisconsin Division of Health declared that nine of twelve teachers in that state who had developed the disease had been bushwhacked by asbestos at their schools. The other three had held jobs prior to teaching—on an iron-ore ship, in construction—that could have contributed to their exposures. In one of the closing sessions, Selikoff took aim at an industry-backed scientist. Instead of sinking more money into studies, as the scientist had recommended, "we should . . . use it to remove asbestos where required," Selikoff said tartly. Barry Castleman attended the proceeding and understood its significance. "Selikoff was disgusted at the idea that there's no hazard from asbestos in place," Castleman said years later. "Calling it the third wave was a frontal assault on this propaganda exercise."

A quarter-century after the New York conference, I encountered one of the faces of the third wave. He was Kris Penny, a fit-looking, thirty-nine-year-old owner of a flooring company in Florida. When I met him at the University of Maryland Medical Center in April of 2015, he was suffering from peritoneal mesothelioma, a rare cancer of the lining of the abdomen almost always tied to asbestos exposure. He'd felt a stabbing pain in his gut after drinking orange juice at a McDonald's one morning in September 2014. Emergency surgery exposed a belly full of tumors. When he awakened, his wife, Lori McNamara, was crying. The surgeon told him to get his affairs in order. Penny had inhaled the microscopic asbestos fibers a decade earlier while laying fiber-optic cable underground in asbestos cement pipes, putting him at the lower end of the latency period. I watched his body melt away in the ensuing months—mesothelioma devours its victims from the inside, until they are cadaverous—and he died at forty in April of 2016. His cancer, like the cancers at Goodyear in Niagara Falls, was sadly predictable. If workers aren't warned explicitly about hazards—and Penny said he wasn't—they'll endure almost anything for a paycheck, even mouthfuls of white dust.

Equally foreseeable was the resurgence of silicosis, an incurable lung disease caused by the inhalation of microscopic particles of quartz. It was the cause of a major occupational disease scandal in the early 1930s: the Hawk's Nest Tunnel episode, in which at least 764 laborers, most of them Black, died after drilling a water tunnel into silica-rich rock near Gauley Bridge, West Virginia. Silicosis became the nation's biggest industrial menace and was the subject of congressional hearings in 1936. In the opening session, Philippa Allen, a social worker from New York who had spent the previous four summers in West Virginia, gave this account:

> These were robust, hard-muscled workmen, and yet many of them began dying almost as soon as the work on the tunnel started. With every breath they were breathing a massive dose of silica dust. . . . Every worker examined by a physician after working in the tunnel any length of time has been found to have this dreadful disease. It is a lung disease that cannot be arrested once it is started. Ultimately, the victim strangles to death.

Allen was introduced to a worker named George Houston, twenty-three, who was "walking very slowly and breathing with effort." He was close to death, having shoveled muck and operated a drill in the tunnel's No. 1 heading for only forty-eight weeks. When he climbed stairs, he told Allen, "it gets me to breathing so hard I have to lay down." The workers, stuffed into hovels at night, made thirty cents an hour and were forced to pay seventy-five cents a week for a doctor who never came to see them. When they died, one local resident said, "they buried them like they were burying hogs, putting two or three of them in a hole." Labor Secretary Frances Perkins organized a national conference and commissioned a public-service film, *Stop Silicosis*, to call attention to the crisis. The disease faded into the background, never going away entirely, only to return with a vengeance among sandblasters in the 1980s and 1990s, coal miners in the early 2000s, and fabricators of artificial-stone countertops in the 2010s. The young ages of the victims in all three groups suggested these were extreme exposures. One fall morning in 2019, I met Fernando Salmeron, the forty-six-year-old owner of a small fabrication shop in the Bay Area city of Benicia, California, whose lungs were full of pulverized sand from cutting and grinding countertops. A native of Michoacán, Mexico, he'd gotten into the business nineteen years earlier to escape the drudgery of strawberry picking, working at someone else's shop before opening his own. Sales were strong; the fake stone, a mashup of crushed quartz and plastic, is cheaper and more durable than natural stone and can be colored to the customer's liking. But it has a far higher crystalline silica content—upward of 90 percent—than granite or marble. For a long time, Salmeron said, he knew little about the toxicity of the white powder that floated through the air and settled in his chest. By the time I saw him he'd lost thirty-five pounds, was short of breath, and was awaiting a double-lung transplant. "I've been at home, doing nothing," he complained. The desk and the Apple computer in his office were covered in dust. (Salmeron had his transplant in April 2022; this will extend his life by, at most, ten years.)

That evening, in Hayward, I met siblings from El Salvador, Rodrigo and Dora Alicia Martinez, who had lost two brothers to silicosis. The brothers, Rafael and Margarito, had worked at a fabrication shop called Stone Etc. and died two months apart in 2018. Rafael, the older of the two, had been sick for ten years; Margarito fell ill more quickly and expired before Rafael. Their deaths were investigated by California's Division of Occupational Safety and Health, which proposed nearly $900,000 in fines against Stone Etc. for violations at the company's two locations, in Hayward and Gardena. The company was still contesting those citations at the end of 2022. The deaths were included in the Centers for Disease Control and Prevention's count of eighteen silicosis cases in four states. As of 2018, the CDC said, there were 8,694 fabrication sites with 96,366 employees in the United States—an average of twelve workers per site. Owners of these operations tended to know little about dust control—or not share what they did know. They were staffed mostly by young Latino men—immigrants who were unlikely to complain. Such was the case with the Martinez brothers in Hayward, who were reluctant to let go of a job that paid fourteen dollars an hour, even though it was killing them. They were buried beside one another in the village of El Carmen, El Salvador. I asked Rodrigo if he believed Stone Etc. knew its workers were being poisoned. "They knew everything," he said in Spanish. Dora wiped her eyes with her blue striped shirt.

I'd been introduced to Salmeron and the Martinez siblings by Robert Harrison, an occupational physician in San Francisco. Harrison was born on September 14, 1954, and grew up in Freeport, Long Island, not far from Jones Beach. He was the second son of Harry and Helen Harrison, both of whom were members of the Communist Party USA. He remembers his father—a frustrated ladies' garment salesman and, later, a junior high school science teacher—bringing home the *Daily Worker* in a brown paper wrapper. His mother taught education at Long Island University. From an early age, Bob was conditioned to empathize with the ordinary men and women who kept the American economy running. Both of his parents were union leaders who participated in civil rights and anti-nuclear demonstrations in the 1960s and '70s. He read Karl Marx's *Das Kapital* at the urging of his older brother, Fred, in high school, and developed an appreciation for labor history, in which he majored as an undergraduate at the University of Rochester. His studies and his father's life experience—he was a "thwarted doctor" who'd been denied entry to medical school in the 1930s because he was Jewish, Bob says—steered him toward a career in occupational medicine. He enrolled in the Albert Einstein College of Medicine in the Bronx in the fall of 1975. His professors there included Victor Sidel and H. Jack Geiger, among the founders of Physicians for Social Responsibility, whose affiliate, International Physicians for the Prevention of Nuclear War, shared the Nobel Peace Prize in 1985. Geiger also helped form Physicians for Human Rights, which shared the 1997 prize for its efforts to ban land mines.

Both Sidel and Geiger were champions of social medicine, which holds that poverty, unsafe working conditions, and other social factors must be considered if physicians want to get at the roots of illness. With grant money from Lyndon Johnson's Office of Economic Opportunity and Tufts University, Geiger and two other doctors set up the nation's first community health center in the mostly Black, desperately poor Mississippi delta town of Mound Bayou in 1966. The clinic, modeled after one Geiger had seen in South Africa, famously handed out "prescriptions" for food. Recipients "would take the food order to the grocery store, which would bill the community health center, and we'd pay for it from the pharmacy budget," Geiger told *Think*, an online publication of Case Western Reserve University, where he attended medical school. "That led to this iconic exchange: The governor of Mississippi screamed at someone in the poverty program, who came down and screamed at me. 'What in God's name do you think you're doing giving away free food and charging it to the pharmacy? A pharmacy is for drugs to treat a disease.' And I said, 'The last time I looked at my textbooks, the most specific therapy for malnutrition was food.' And so he went away because he couldn't think of anything to say to that."

At Einstein, Sidel brought in guest speakers to expose the students to a range of thought outside of traditional medicine. One of them was Tony Mazzocchi, who made such an impression on Bob Harrison that he took an internship with the OCAW in the summer of 1976, joining three other medical students from Einstein and a graduate student in economics from the University of Massachusetts to investigate high rates of disease at the Merck Pharmaceuticals plant in Rahway, New Jersey. Workers there were suffering from, among other things, asbestosis and elevated serum enzymes, indicative of liver damage. The students' report to the union, *Merck Is Not a Candy Factory* (the sardonic line bosses used on complaining employees), chronicled in impressive detail the plant's inner workings. "Exposure to chemicals is a severe problem for both operators and mechanics," the students wrote. "In manually charging a still, taking a batch sample, digging out a non-automated centrifuge, feeding a drier or packaging the product, and operator inhales chemical vapors, and literally eats the chemical powders." Respirators were worn "infrequently" and offered "little, if any, protection against the prolonged and extensive exposure associated with improper maintenance and feeble ventilation." Solvents, including benzene, were being stored in open containers. Workers in one part of the plant were subjected continuously to chloroacetone, a pungent liquid used as a tear gas during World War I, resulting in "frequent gassings and occasional loss of consciousness." Fumes escaped the plant and ate away the paint on cars and homes in the neighborhood; Merck had to pay for repainting.

"We did not discover patterns of occupational disease as dramatic as polyvinyl chloride-induced angiosarcoma or asbestos-induced mesothelioma," the students wrote. "But, the long list of chemical and dust exposures, physical hazards, and disease contracted at the Merck Pharmaceutical plant are no less appalling. We recognize the need of people to provide for themselves and their families, and our study is not intended in any way to threaten the continued ability of people to do so. However, we cannot accept that working people must earn a living by sacrificing years of their lives."

This thinking—your job needn't take your life—guided Harrison after he graduated from medical school in 1979 and moved west to be with his future wife, Robin, who was in training as a psychiatrist. He helped launch a worker clinic at San Francisco General Hospital and did a three-year residency in internal medicine at the University of California, San Francisco's Medical Center at Mount Zion, followed by a two-year residency in occupational medicine, also at UCSF. He joined the university's faculty in 1984 and began tracking worker illnesses and injuries for the California Department of Public Health. Over the next few decades he saw thousands of patients: industrial painters, shipyard laborers, Silicon Valley technicians. He coauthored papers on methylene chloride–induced asphyxia among bathtub refinishers, workplace violence, and HIV among male adult-film actors. In the spring of 2022, he learned of another silicosis cluster, this one in Southern California. Since January 2016, twenty-five artificial-countertop workers—again, all relatively young Latino men—had been diagnosed at Olive View–UCLA Medical Center in the San Fernando Valley with a severe form of the illness; at least five more workers in the Los Angeles area had been diagnosed elsewhere. I met two of the workers that fall. Both Gustavo Reyes Gonzalez, thirty-two, and Juan Gonzalez Morin, thirty-six, were emaciated and hooked up to portable oxygen tanks. Both hoped to qualify for lung transplants. But neither would reach old age under the best of circumstances. Jane Fazio, a pulmonary physician at Olive View, said each had a type of lung scarring known as progressive massive fibrosis, which inflicts damage even after exposure stops. She and her boss, Nader Kamangar, believed they'd found the biggest cluster of silicosis among fabrication workers in the country. Kamangar pulled up an image of a victim's CT scan on his computer. Healthy lungs, he explained, should appear dark. These looked mostly white, indicating widespread destruction of tissue. Without a transplant, the patient probably would live no more than a year. "I have spoken to God," Reyes said in Spanish during my interview with him. "If it is my time to go, I am happy to go with him. . . . If the transplant comes first or it doesn't come, I have accepted it." Reyes got his transplant in February 2023, had gained weight, and was even jogging. Gonzalez's body gave out, and he died on April 9—Easter Sunday. He had just turned thirty-seven.

As Harrison saw it, the countertop manufacturers were largely responsible for this flare-up of a very old disease. "They're making silicosis in a box and shipping it out," he said. "How could they not know?" How, indeed? In 1938, the Labor Department released *Stop Silicosis* to address the "wave of fear" that had emerged in American workplaces. Narrator: "Cause of the disease: dust. Results of the disease: disablement, poverty, death. Cure for the disease: none." Risky jobs are portrayed: granite quarrying, sandblasting, metal casting. We meet the luckless "John Steele," who works in a foundry where dust is uncontrolled. He becomes so "sick and incompetent" that he's let go. "He is paid off, broken in health and in spirit." Steele can find work only in a tombstone shed, where he etches his own name onto a slab. The narrator's tone lightens as he explains that there are solutions to the crisis: wet drilling, supplied-air masks, high-efficiency exhaust systems. And yet, more than eighty years after the film was released, workers in California were still cutting and grinding silica-laden slabs in garage-size shops, generating clouds of what looked like powdered sugar. The particles, one hundred times smaller than a grain of sand, went deep into their lungs and sealed their fate.

CHAPTER 20

KIDS

Having seen the trauma their newborn children's birth defects still inflicted on Ray and Dottie Kline, I also wanted to know more about links between parental exposures in the workplace and fetal harm. This led me to Mark Flores. When I met him in the summer of 2014, he was watching *Sesame Street* in the living room of his mobile home in San Jose, California. The show held his attention like no other, which wouldn't have been surprising if Mark had been four. In fact, he was thirty-four. He stared absently. He mumbled. He tended to repeat whatever his mother, Yvette, had said last, and sporadically uttered complete sentences. Toy trucks, tractors, and dolls were scattered about. Yvette had worked in the electronics industry as a young woman from 1975 to 1979, in the early days of Silicon Valley. She made barcode scanners for a company in Mountain View called Spectra-Physics, working in a stuffy room in which she fused laser tubes with a paste she'd made by mixing a green powder with a clear liquid. She earned $2.70 an hour and wore a paper mask, which was all but useless against the dust and fumes. The green powder, she'd learn many years later, was 62 percent lead. The clear liquid was methanol—wood alcohol, best known as a contaminant in moonshine.

Both lead and methanol are very bad for fetuses. Yvette, a young woman happy to have a steady job and the independence that came with it, had no inkling. And so, even after she miscarried in a plant restroom

170

in 1978, she went back to work, unprotected. Early the following year, Yvette became pregnant with Mark. He was born on December 3, 1979, with macrocephaly (an enlarged head), crossed eyes, blood blisters on his head, undescended testicles, and dislocated hips. He was still just crawling at four, unable to form words. His parents bribed him with food to get him to talk, and by the time I met him he weighed nearly four hundred pounds. His father, David, a truck driver, died in a motorcycle accident in 2007. That left Yvette to care for Mark, with occasional help from his younger brother and sister. Their life consisted of visits to doctors and adult day care and outings to Target. Yvette, in frail health herself, would lapse into bouts of exhaustion and depression but always rallied.

Yvette assumed that Mark's profound mental impairment was simply bad luck until she heard a radio ad from a law firm, seeking electronics workers whose offspring were born with defects. Yvette called the number in the ad, provided some basic information, and soon heard from Amanda Hawes, a diminutive lawyer in San Jose who specialized in representing women who'd done the dirtiest, most dangerous jobs in the canning and electronics industries. Hawes set about learning what Yvette might have been exposed to at Spectra-Physics that did such dreadful things to Mark. It took her four years and considerable detective work to determine that the likely agents were lead and methanol. Yvette's lawsuit against Spectra-Physics was settled in 2013; the company admitted no liability.

In November 2021, Mark and Yvette finally left the San Jose trailer park—too noisy and dangerous, Yvette said—and moved in with Yvette's daughter, Lisa Marie, who had a two-story house in an exurb in California's Central Valley. Mark had been allotted three rooms for his toys and could roam freely in the fenced backyard. When I spoke with Yvette in May 2022, she said Mark's weight had fallen to 220 pounds and he'd made progress in other ways: "He now knows how to pour water into cups without spilling it," Yvette said. "He's learning to say no at the appropriate time." He picked out his clothes and assigned names to his stuffed animals. At forty-two, he still insisted on sleeping in the same room as his mother, however, and still threw things in anger when he was feeling melancholy about his late father. "He doesn't know how to verbalize," Yvette explained. She was taking chemo pills for a form of blood cancer known as essential thrombocythemia, which sapped her energy, gave her headaches, and made her bruise easily. She wondered if this, too, might have been related to her work at Spectra-Physics.

At the same time Yvette was being forcibly served the lead-methanol cocktail that likely short-circuited Mark's brain, LeAnn Severson was sorting silicon wafers in a factory in Mountain View owned by NEC Electronics. She remembers working, gloveless and without respiratory protection, with a liquid that smelled like fingernail-polish remover. It was methyl ethyl ketone, MEK, one of several solvents that workers used to clean the wafers. "We had this little thing, like a salad spinner, to hold the wafers down with suction," LeAnn told me by phone. "The idea was to spin them dry. It [the MEK] would blow in our faces." Darryl Severson was born in July 1980, seven months after Mark Flores. His head, unlike Mark's, was abnormally small, but he otherwise followed Mark's developmental path. In his forties he still watches cartoons obsessively, is fascinated by clocks, CDs, record albums, and wristwatches, and, as his mother put it, has "the mentality of a small child." I heard him in the background as LeAnn spoke, and he had the same clipped delivery—"yeah, yeah"—I'd heard from Mark. LeAnn, who lives in Spokane, Washington, sent me a recent disability assessment from a caseworker with the Washington State Department of Social Health Services. She thought it was an accurate portrayal of Darryl's situation.

> Darryl is very social. He will talk a great deal when excited and may need direction to stop. . . . Darryl can get excited and talk too fast, he can get keyed up and has to be monitored for coughing when he gets this way. . . . [He is] fascinated by babies. . . . Darryl doesn't know boundaries. . . . He likes to be reassured that everything is okay. . . . Someone must be with him so he is not taken advantage of. . . . He has never indicated an interest in any romantic relationship.

Like Yvette Flores, LeAnn was represented by Mandy Hawes and reached a settlement with her former employer in Silicon Valley and a chemical supplier. Like Mark Flores, Darryl is physically strong and mentally frail. He will always need help bathing and brushing his teeth, and can't be left alone outside of his house. The chemical company tried to blame Darryl's disability on Fragile X syndrome, a genetic condition, but testing revealed no such defect, LeAnn said. She'd long ago had a "gut feeling" that her exposures at work had done the damage.

Dr. Cynthia Bearer, a neonatologist in Cleveland, was a plaintiff's expert in both the Flores and the Severson cases. The chemicals to which Yvette Flores and LeAnn Severson were exposed can enter the placenta and wind up in the fetus, reacting with DNA or cellular signals to disrupt development, Bearer told me. The results can range from fetal death—recall Yvette's miscarriage in the factory restroom—to structural or functional defects. Regulatory agencies haven't responded well to the problem. "People want to see bodies," Bearer explained. Had thalidomide, a morning-sickness drug administered in the 1950s and '60s, caused only mental retardation instead of malformed limbs, "it might not have been noticed," she said.

As was the case with DuPont's bladder-attacking dyes, knowledge of teratogens—agents that can cause fetal abnormalities—in the workplace had entered the scientific mainstream by the nineteenth century. In a study published in 1860, French physician Constantin Paul followed the pregnancies of seven women married to workers who'd shown evidence of lead poisoning. Between them the women had thirty-two pregnancies. Of these, eleven ended in miscarriage and one in stillbirth. Of the twenty children born alive, Harvard's Alice Hamilton reported in *Industrial Poisons in the United States*, "eight died in the first year, four in the second, five in the third, and one after the third, leaving only two alive out of 32 pregnancies."

In his portentous 1911 lecture, Sir Thomas Oliver spoke of how he'd helped secure "the emancipation of female labour from the dangerous processes of lead-making. . . . Lead hits hard the reproductive powers of man and woman, but especially of woman." Papers published in 1936 and 1937 described the narcotic and neurotoxic effects of solvents used to produce newly popular "fused" shirt collars, which remained stiff without the use of starch. One female worker in Ohio became so drowsy and disoriented that she had to be admitted to a psychiatric hospital. She was "confused and spoke thickly, slowly and jerkily," a resident at Cincinnati General Hospital reported. "At times she was completely disoriented and thought she saw or heard her daughter in the ward. If undisturbed, she dropped off into sleep." Two of nineteen young male workers studied in New York State became acutely ill, presenting "a picture of partial narcosis with marked tremor of the hands," industrial hygienists with the state Department of Labor wrote. "When spoken to, they did not seem to realize that they were being addressed and responded with considerable hesitation and halting after a prolonged reaction time. Their conversation was sluggish and enunciation indistinct. There was a question in the minds of the examiners as to whether or not the boys were subnormal mentally." In both cases suspicion fell on Methyl Cellosolve, which belonged to a family of chemicals known as ethylene glycol ethers, later used in electronics manufacturing and implicated in spontaneous abortions and birth defects.

By the beginning of World War II, as women entered the American workforce by the millions, it was clear that those of childbearing age required special consideration. In 1942, the federal Labor Department published *Standards for Maternity Care and Employment of Mothers in Industry*. A recommendation: pregnant women should be kept away from substances like lead, mercury, and benzene. And a warning: "Because these substances may exert a harmful influence on the course of the pregnancy, may lead to its premature termination or may injure the fetus, the maintenance of air concentrations within the so-called maximum permissible limits of state codes is not, in itself, sufficient assurance of a safe working condition for the pregnant woman." Two years later the War Department published a technical bulletin detailing the hazards of industrial solvents, including menorrhagia—prolonged menstrual bleeding—in women. Inhalation and absorption of the chemicals should be avoided, the bulletin advised, and less toxic products should be used whenever possible. In 1946, the National Research Council offered insight into the burden pregnancy imposed on "every organ" in a woman's body. "The ability of the tissues to balance injury with repair may be considerably altered. It would seem that any chemical substance which is capable of producing a harmful effect on the internal systems of the body would be of greater danger under these conditions." Pregnant women, the council said, "should not be allowed to work at occupations involving exposure to harmful chemical substances which produce systemic damage, anoxemia [an absence of oxygen in arterial blood], irritation of the respiratory tract, and the like. Because these substances may affect adversely the pregnant woman or the fetus, the concentrations usually accepted as allowable should not be considered safe for pregnant women."

By the time Yvette Flores began working at Spectra-Physics in 1975, the health risks associated with the assembly of printed circuit boards—the substrata of electronic devices on which semiconductors, diodes, resistors, and capacitors were mounted—had become evident. In a frank 1972 talk, published in the *Annals of Occupational Hygiene*, British scientist W. MacLeod Ross observed that solvents and chemical cleaning agents were "often handled with an abandon which horrifies a trained chemist or safety engineer." So-called clean rooms—designed to protect the product and not the worker—were repositories of filthy air. Ross cited one workplace, which he didn't name, where only 20 percent of the air workers breathed was fresh; the other 80 percent was recycled. "Despite the obvious precautions of extraction units to remove solvents from drying resists, small quantities escaped and gradually produced a concentration high enough to cause acute operator discomfort with a risk of chronic damage," he said. The employer solved the problem by installing activated charcoal filters in the recirculation system. Ross worried, however, that while workers generally knew to avoid poisonous salts and acids and alkalis that could dissolve tissue, "many are brain-washed into accepting solvents of various types as 'safe.'"

In 1976, Yvette's second year of building lasers, the Conference on Women and the Workplace, co-chaired by Eula Bingham, was convened in Washington. It was trailblazing and at times testy. Some female attendees recoiled at what they believed to be patronizing remarks by their male counterparts; twenty-three of them issued a declaration, mid-conference, objecting to being seen as "reproductive vessels. Both men and women are involved in reproduction. Reproduction is not solely the woman's responsibility. Society is ultimately responsible for the preservation of future generations." They also objected to "paternalistic statements which assume that if 'you' women workers are educated about job hazards, then 'you' women can exercise free choice about 'your' employment. Given the present unemployment rate and discrimination against women workers, this is an absurd and condescending attitude toward both women and men workers."

Among the conference speakers was Dr. Josef Warkany, an Austrian-born pediatrician who became a professor at the University of Cincinnati College of Medicine and was among the first researchers to tie environmental exposures to birth defects. Warkany opened his presentation by saying that the medical literature of the nineteenth century was "replete with statements which blame children's defects on the father's or mother's alcoholism, tuberculosis or syphilis. . . . Many of these correlations were accidental and not causative [and] could not stand analysis in later years." Experiments were performed on the eggs of chickens, amphibians, and fish, but these tests had little relevance to humans. "It was thought that the human embryo and fetus were so well protected by the mother within the uterus that they could not be deformed by environmental factors," Warkany said. By the 1930s, "it seemed more 'scientific' to consider most, if not all congenital malformations as genetic and hereditary." This was due in part to the lack of animal evidence; the field of experimental teratology materialized shortly thereafter, first drawing a connection between defects and dietary deficiencies, starting with vitamin A in rats. "These experiments were, and are, of great theoretical interest," Warkany said. "They proved, once and for all, that environmental hardships imposed on pregnant females could cause structural malformations in the young." Still, he cautioned, many of the findings didn't translate to humans. And the "best-known teratogen in man," thalidomide, was shown to inflict injury on monkeys and rabbits only after the fact.

The three-day meeting featured traditional presentations and livelier discussion periods. The medical director of Dow Chemical, Harold Gordon, fretted that his company could be held liable for discrimination if it excluded fertile women from certain high-risk jobs, and for harm to their offspring if it didn't. Dow was trying to do the right thing, Gordon said, but there were many questions: "What is the population at risk? How is it identified? What specific agents need to be considered? How are they identified? What restrictions are needed? When do they apply? To whom do they apply?" The OCAW's Tony Mazzocchi was unsympathetic. "There isn't a single worker, including those at Dow, who knows what the hell he works with in the first place," he said. "No industry has allowed us to look at monitoring data. . . . We aren't being told what is carcinogenic or what is teratogenic. We are learning after the fact." It was the union's position, he said, that neither women nor men should have to be removed from a job to protect the unborn; "our position is to make the workplace safe for everyone." Dr. Bertram Carnow, a professor at the University of Illinois School of Public Health, agreed in principle. If it turned out that male workers were as responsible for birth defects as female ones, "then we also have to remove the men [from exposures]," he said. "And we'll either wind up with 60-year-old eunuchs in the workplace or we'll have to eliminate the teratogens. I suggest it would be easier to eliminate the teratogens."

For workers like Yvette Flores, the most germane speaker was Andrea Hricko with the Labor Occupational Health Program at the University of California, Berkeley. Hricko had been investigating, with growing alarm, the electronics industry, which then employed some 271,000 workers in the United States. More than three-quarters of these workers were women, she said, and many were "nonwhite. A high percentage do not speak English. And few of them are organized." Hricko said she'd heard from some California workers that "dark-skinned women, including Filipinos, Chicanos and Blacks, are often assigned to certain departments in electronics plants where chemicals are used that cause skin rashes. This is done because the skin rashes are not as obvious on the dark skin and, therefore, the workers don't complain as much and the company doesn't have to deal with the complaints." Some plants didn't allow languages other than English to be spoken, "even in departments where the majority of women are non-English-speaking," Hricko said. Intimidation was common. Women in one Silicon Valley plant who'd prepared a warning leaflet on trichloroethylene "had to leave copies of it in the women's rest room so that the male management people would not snatch the copies away."

Eight days before the conference began, Mandy Hawes gave birth to her first child, Gordon. Hawes, then thirty-two, had been born and raised in Glen Rock, New Jersey, the second of four children. Her father, John, was a banker with a progressive bent; her mother, Judith, a housewife and, later, a special-education teacher and an author of children's books. Her youngest sister, Lester Anne, was born with Down syndrome in 1952 and sent to a "training center," an institutional setting popular at the time, at age two. Hawes attended Wellesley College, where she fell under the influence of a Spanish literature professor named Justina Ruiz de Conde, who as a young lawyer had gone to bat for displaced children of Republican fighters during the Spanish Civil War and fled to the United States when the Republic fell to Francisco Franco's Nationalist forces. Ruiz's energy and earnestness—she'd lobbied Eleanor Roosevelt in New York to arrange safe havens for the children during the war—motivated Hawes to attend Harvard Law School. After graduating in 1968, she moved with her then-husband to Berkeley and began working as a legal-aid lawyer in San Francisco and then Oakland, where she helped women working in East Bay and San Jose fruit and vegetable canneries secure back pay that had been illegally withheld. Hawes learned of the barbaric working conditions the women faced. The factories were hot, wet, and dirty. Fast-moving conveyor belts sliced the skin and crippled the hands and wrists. Promotions were sometimes contingent on unwanted backseat trysts with supervisors.

By 1977, when Hawes went into private practice, most of the canneries had relocated to California's Central Valley or Mexico. In their place came the semiconductor factories, which tapped the same, mostly female and minority, workforce. Hawes's first three clients from the industry worked with no respiratory protection in the R&D laboratory at Signetics in Sunnyvale, where, they said, they'd been sickened by exposures to solvents, acids, and heavy metals. Their throats, noses, and tongues burned; they had recurrent headaches, lightheadedness, and giddiness (not unlike the tipsy PVC reactor cleaners at Goodyear in Niagara Falls). When they complained, they were reassigned to the cafeteria, where they were given nothing to do. They were fired after Signetics insisted it couldn't find other work for them. A NIOSH evaluation of the plant, requested by Hawes, confirmed the workers' allegations: investigators collected thirty-three air samples and detected solvents including toluene and xylene but at levels "well below" what NIOSH recommended and the state of California permitted. "In [the investigators'] opinion the problem was caused by one or more chemical agents in certain areas of the Signetics workplace, which, upon becoming airborne, are capable of irritating the mucous membranes and inducing an altered state of response in some persons," NIOSH found. "Also, certain symptoms consistently reported by Signetics employees suggested the possibility of intermittent exposures to a narcosis-producing agent. Because of the restrictions placed on the investigation [by the company], the true population at risk and the true population affected could not be satisfactorily identified." The investigators were "of the opinion that a significant occupationally-related health problem exists at the Signetics Sunnyvale facility" and urged that a "larger, more systematic" study be undertaken. Hawes filed workers' compensation claims and a lawsuit on the ousted Signetics workers' behalf; all were settled.

In 1988, epidemiologists at the University of Massachusetts reported finding twofold elevations in miscarriages among women who'd performed certain tasks at the Digital Equipment Corporation factory in Hudson, Massachusetts. The sample size was small, however, and the scientists proposed a study with a larger population. That came in the 1990s, when the Semiconductor Industry Association hired a team from the University of California, Berkeley, School of Public Health to do a retrospective study of workers at fourteen manufacturers. One of the researchers, Katharine Hammond, went into the project expecting to find nothing of consequence. Instead, the team found high miscarriage rates among women who'd handled photoresist (a light-sensitive resin), ethylene glycol ethers and other solvents, and fluoride compounds. The risk, relative to the general population, was more than threefold for workers who'd been exposed to all these substances.

In the late 1990s, Hawes joined other lawyers who'd begun suing IBM. A testicular cancer cluster had emerged among male workers at the company's plant in East Fishkill, New York. Female employees were developing kidney cancer, leukemia, and lymphoma at higher-than-expected rates, and workers' children were being born with defects such as epidermolysis bullosa, a whole-body blistering of the skin that tortured twins Kate and Kelly Daley all of their twenty-seven years before they died thirteen days apart in 2006. Their father, Chris, a technician in East Fishkill, succumbed to non-Hodgkin lymphoma the following year. "Initially, it was a group of guys with testicular cancer, a ridiculously large number, just an improbable cluster," one member of the legal team, Steven Phillips, recalled. "It was rapidly turning into birth-defect litigation because as we started investigating, all these birth defects came jumping out. They were so severe and unusual, and there were so many of them. It was a catastrophe."

After the IBM cases were resolved, Phillips and the team turned their attention to Motorola, which had semiconductor operations in Arizona and Texas. By early 2023 they'd brought more than eighty birth-defects cases against the company, about three-quarters of which had been settled. In a memorandum for punitive damages filed in 2018, the lawyers accused the company of abiding conditions it knew posed "severe reproductive dangers." The forty-six-page document outlines the state of knowledge about teratogenic hazards in the workplace decades before the children were born. "The basic biology of shared blood between the mother and the fetus was well known, as was the capacity of many organic solvents to cross the placental barrier," the memo says. "The immature immune system of the developing fetus was particularly susceptible to toxic chemical exposures. This was textbook medicine by the 1940s."

What became Motorola was founded in Chicago in 1928 as the Galvin Manufacturing Corporation, named after brothers Joseph and Paul Galvin. It made its mark selling an automobile radio called the Motorola, then expanded into making tabletop radios for the home and handhelds for the military—walkie-talkies, widely used during World War II. The company changed its name to Motorola in 1947, began producing low-cost televisions the following year, and by 1956 had established its Semiconductor Products Division in Phoenix. In 1974 it sold its Quasar color TV line to a Japanese company so it could concentrate on making microprocessors for computers. Here, as Phillips, Hawes, and their colleagues tell it, is where the trouble began.

Janice Numkena, a member of the Hopi tribe, began working in the clean room at Motorola's Fifty-Second Street plant in Phoenix in 1973, transferred to another Motorola plant in Mesa seven years later, and left the company in 1998. She gave birth in 1976 to a daughter, Heather, who suffered from hearing loss and developmental disabilities. Six years later, another daughter, Angela, was born with cerebral palsy and was never able to walk, speak, dress, or feed herself. "She didn't have a childhood," her father, Robert, testified in a 2014 deposition. "She never played with dolls. She couldn't run around. She didn't have friends. She didn't have a pet. And just the normal things that little girls do, she never had that."

Erma Acosta started at the Fifty-Second Street plant in 1968 and moved to Mesa in 1970. She bore a son, Michael Santa Cruz, two years later. She worked in wafer processing and, she testified in a 2013 deposition, received no warnings about the chemicals she used: photoresist, hydrofluoric acid, sulfuric acid, benzene, trichloroethylene, xylene, acetone, freon. She wore gloves only when handling the acids, she said, and vied with the other female workers to grab the driest gown with the fewest holes at the start of her shift. The work area "smelled like [nail] polish remover," Acosta recalled, and exhaust fans worked only "occasionally." She suffered headaches and nausea daily. The latter was different from the morning sickness she'd experienced during an earlier pregnancy, she explained. Those bouts of nausea "would subside. But when I was pregnant with Mike, the nausea seemed to just be ongoing."

Michael was born on February 22, 1972, with a condition the doctor described as myelomeningocele. "And I said, 'What's that?'" Acosta testified. "He said, 'It's when they're born with their spine outside of their body.'" The doctor explained that there were degrees of severity and that Michael was at the upper limit. He told Acosta some protrusions were the size of a pea; her son's was the size of a grapefruit. Acosta asked the nurses if she could see him. "And they said, 'Well, we have him . . . like, in a storeroom, and we're just waiting for him to pass.' And I said,

'What?'" Acosta took a shower, collected herself, and learned from one of the nurses that Phoenix's Crippled Children's Hospital accepted patients with severe defects. "I said, 'Well, where is it?' And she gave me the address." Acosta asked for some bandages in which to wrap the little boy, who was "a bloody mess . . . but he was so cute." Doctors at Crippled Children's saved Michael's life, but he underwent many surgeries and was paralyzed from the chest down. In his forties he still read at a third-grade level, wore diapers, and was easily disoriented.

Acosta delivered three other children—one older than Michael, two younger—with no defects. Her husband at the time of Michael's birth had been the first to connect her workplace exposures to the infant's plight. "He had stated to me that the chemicals that I was working with were highly toxic, and that he couldn't believe that . . . we as a group were working with them," Acosta testified.

As a founding member of the Semiconductor Industry Association in 1977, Motorola "had access to state-of-the-art research and information about health hazards posed by its manufacturing chemicals and processes," the plaintiffs' punitive-damages memo says. In 1981, Peter Orris, then a NIOSH regional medical director, was invited to speak at a seminar on female employees' health at Motorola headquarters in Schaumburg, Illinois. Orris "emphasized to Motorola management that pregnant women should not be exposed to solvents and lead," the memo says. He recommended reassignment as an easy solution to a problem that "could not be solved by engineering or protective gear alone." The same year, DuPont sent a letter to Motorola and other purchasers of ethylene glycol ethers, cautioning that "employee exposures should be reduced to the practical minimum." Motorola understood the significance of this warning but did not tell its fabrication workers, according to the memo. Nor did it adopt "a scientifically and medically sound reproductive health policy and a comprehensive and reliable industrial hygiene monitoring program for chemicals at issue." Like Goodyear, it insisted that if inhalation exposures were within federal limits, its workers were safe. "This course may have spared Motorola an OSHA citation, but it did little to protect the employee and her fetus from the risk of regular exposure at reproductively toxic levels," the memo says.

The company's indifference, the plaintiffs allege, led to the malformations that afflict Angela Numkena, Michael Santa Cruz, and children whose mothers were exposed to teratogens at work as recently as the early 2000s. In its response, Motorola said that "the evidence confirms that Motorola did not, at the time these Plaintiffs were *in utero*, have knowledge or notice that any of the chemicals used posed reproductive or teratogen risks, including the various specific injuries alleged here. That is hardly surprising, given that whether such risks even exist in the first place remains, decades later, a matter of scientific debate."

In the early 2000s, semiconductor manufacturing began moving overseas, to countries such as Taiwan, Vietnam, and South Korea. But the United States is trying to lure it back with generous subsidies. By the summer of 2022, Mandy Hawes had devised a plan to hold the chipmakers and electronics assembly facilities accountable for their transgressions in Silicon Valley, beyond what they'd been handing out in settlements of individual cases. She was sensitive to the tribulations of the mentally impaired—her younger sister with Down syndrome had been shunned by much of society until her death at fifty-nine—and she knew well that the challenges for caretakers made for an exhausting, lifelong commitment; people like Mark Flores don't suddenly become independent. Hawes hoped to persuade researchers, public education administrators, and elected officials to investigate whether a disproportionate number of children placed in special education in Santa Clara County due to developmental disability—and in a largely custodial state-run program once they "aged out" of special ed—had mothers who'd worked in the electronics industry when pregnant. (This could be determined by analyzing parental occupation data recorded on birth certificates since 1980.) To the extent it was responsible, Hawes reasoned, the industry should help offset the high costs incurred by the public to provide these services, just as opioid manufacturers were being forced to pay for the health crises they'd fueled.

CHAPTER 21

RAY AND HARRY
IN RETIREMENT

I N SEPTEMBER OF 2017, I made the second of five trips to Niagara Falls to catch up with Harry and Diane Weist. One morning they invited a group of Goodyear old-timers to meet me at the farmhouse in Youngstown. Ed Polka, four months away from retirement, showed up, as did retirees Dickie Dutton and Jack Schoonover. David "Bo" LaClair was in his last months of employment. Four years earlier, I'd written an article about the bladder-cancer cluster, which then stood at fifty-eight cases. Harry was featured in it and took some grief as a result. He was accused of goading Goodyear into closing the plant, a prospect he dismissed as unlikely. Schoonover agreed. "They'd never shut it down because they'd have to dig down three miles to get all the contamination," he said. Harry laughed.

The men at the dining room table—Schoonover was the oldest at eighty, LaClair the youngest at sixty—had worked at the plant in its heyday, kept it profitable, made it collegial. They'd been replaced by a band of joyless ciphers they called the "Children of the Corn," a reference to a 1984 slasher movie in which people over eighteen are targeted for execution by a cult of demon-worshipping youths. Institutional knowledge had dwindled. "They've lost all their experienced millwrights," Harry said. "They brought some people over from production and they brought some people off the street, but it ain't the same animal." "They don't want to learn how to fix shit," LaClair said. "I can't wait to get the hell out of there."

LaClair had joined Goodyear at twenty-two in 1979, bagging Nailax and another rubber chemical, Kagarax, on the midnight shift in Department 245. "It was a ball," he said. The workers had water fights and threw lit cigarettes at one another. Beer would be smuggled in and consumed overnight; the empty cans were carried out in the morning. Once LaClair moved over to maintenance, he and Harry spent many weekends together in the shop—sixteen hours on Saturday, another sixteen on Sunday. The money rolled in. Then, around 2000, Goodyear began hiring youthful "backstabbers" who refused to do certain jobs, LaClair said. "They were all stepping on each other to advance," Schoonover said.

Dutton and Polka had avoided serious illness or injury, but Schoonover's hearing was shot from working near screaming atomizers known as "canaries," and he'd just finished nine months of chemo for prostate cancer. Harry, of course, had bladder cancer. LaClair had blown out both shoulders and said he couldn't "hear shit." Would cancer find him as well? Possibly. He remembered the blue haze that filled the warehouse in Department 245, a "nice cocktail" of the organics used to make Nailax and Kagarax. Before NIOSH came into the plant and Goodyear began tightening up, he'd clean the Sparkler filters wearing a backward ball cap and Salvation Army clothes he'd wash at night in a laundromat. "If you didn't stick your face in and breathe them fumes, you were a puss," LaClair said. "Who wants to be a puss?" All the men at the table were peeved at Goodyear for slashing retirees' medical benefits. Harry had been paying $20 a month for his coverage; now he was paying $400. He and Diane had sold the farm and were preparing to move to Las Vegas for good. Diane was feeling wistful. "I'm just ready to be done with it," Harry said.

Fourteen months after this visit, I spoke with Diane's mother, Dottie Kline, by phone. She said she was a "Christian lady" who had forgiven Goodyear. As the conversation progressed, however, her resentment spilled out. Her husband, Ray, had had two bouts of bladder cancer, a heart attack, and prostate cancer. His rewards for thirty-nine years of service were second-rate health insurance and a gnawing fear that his tumor would reappear. She had lost two children to birth defects she believed were related to Ray's work. "I am a Republican capitalist," Dottie said, "but I don't believe corporations have played fair with their employees, and that's across the board. You cannot put a monetary value on people's lives."

I went back to western New York in July 2019 to attend Guy Mort's workers' compensation hearing in Buffalo and meet two men—Don King and Don Glynn, both eighty-six—who remembered Niagara Falls in the 1950s and 1960s, before it went to seed. King, who met me at the *Niagara Gazette* office on Third Street, ran the high-end Wellesley Dress Shop on Main Street after his uncle died in 1962, later moving it to the Rainbow Centre, a downtown mall that closed in 2005. He served on the Niagara Falls Board of Education for thirty-five years, beginning in 1977, and saw student enrollment plunge. He'd been president of the local Rotary Club and was still a trustee of the public library and Memorial Medical Center. He'd seen the city's population exceed one hundred thousand and drop below fifty thousand. Many of the businesspeople he came to know followed industry out of the city. The chemical plants were loud and smoky and gave off a "wicked smell," King acknowledged, but had been the city's lifeblood. His mother, Polly, an artist, found them captivating and included them in some of her paintings.

King offered a quick driving tour of the city. We went south on Third Street, west on Niagara Street, and southeast on Rainbow Boulevard, passing several new chain hotels on the sparsely developed eastern edge of downtown. We turned onto Buffalo Avenue, where only a few of the factories—Olin, OxyChem—remained. "They had thousands and thousands of people working here," King said. The silos of the old Nabisco plant still stood; King, channeling his mother, thought someone should paint them bright colors and turn them into works of art. We came back into town on the Niagara Scenic Parkway, the Niagara River and Grand Island to our left. "If we'd marketed Niagara Falls right, we'd have had hotels right here along the water," King said. We passed the old Niagara Club, a men-only venue for many years. "Now I don't even know what it is," King said. Here, on our right, was the Rainbow Centre, whose dank parking garage was still being used by tourists, and which housed, on its southern end, the Niagara Falls Culinary Institute and a Barnes & Noble bookstore. The rest of the mall was vacant. In a few minutes we were heading north on Main Street, where King had his dress shop. The building he'd owned between South and Michigan avenues was occupied by a bookstore. "This whole block used to be retail," he said. There were banks, barbershops, florists, furniture stores, shoe stores, hat stores, menswear stores, a department store, a supermarket. One could buy a Pontiac or an Oldsmobile, attend a Presbyterian church, page through books at a classic Carnegie library. None of it was possible now.

King and I met Glynn for lunch. Glynn had become a part-time reporter at the *Gazette* in 1958, went full time in 1960, and retired in 2018. The beginning of his career coincided with the construction of the Niagara Power Project in Lewiston, at the time the largest hydropower facility in the Western world. Built in a remarkable three years, it included two generating plants, two intake structures, two reservoirs, and an assortment of pump stations. Twenty workers had died connecting its six sections, which required the displacement of more than twelve million cubic yards of rock. John F. Kennedy had celebrated the project as a paragon of American grit and ingenuity. Glynn saw it as a "last-ditch effort" to keep industry in the city. In that sense it failed; city boosters who assumed free-spending visitors would make up for the loss of companies like Carborundum and Great Lakes Carbon were mistaken. "People did not take tourism seriously," Glynn said. "They just counted on those people coming every year. Pretty soon, Niagara Falls was no longer the honeymoon capital." It didn't help that downtown looked like a "war zone," an example of urban renewal gone bad, while high-rise hotels sprouted on the Canadian side. The poisoned neighborhood of Love Canal accelerated the city's decline.

Glynn loaned me a 1949 booklet published by Union Carbide. Niagara Falls, it boasted, had built "a chemical and metallurgical industry on which depends the whole economy of the nation. . . . Today we are a city of close to 100,000 people. We are proud of our well-managed city government . . . of our excellent schools and fine university . . . of our many churches and good homes . . . of our park system and well-paved roads . . . and of our health and welfare activities. And, as the 'Power City of the World,' we look to our industries for our continuing growth and prosperity." As it happened, that morning's *Gazette* offered a harsh counterpoint to Carbide's love letter. The newspaper's front page featured an article on Anthony Vilardo, president of USA Niagara Development Corporation. "Looking out over the city . . . the Falls native understands that residents here are looking to the agency to reverse the 1970s and 1980s disaster that was 'urban renewal,'" reporter Rick Pfeiffer wrote. Vilardo and the city administrator, Nick Melson, had recently surveyed the city on foot to see what sort of shape it was in at the height of the tourism season. "We found 46 trees that were missing or dead, 12 [streetlight stands] in need of repairs, four missing bollards at Centennial Circle and a broken tree grate on Rainbow Boulevard," Melson was quoted as saying.

I returned to Niagara Falls two months later to be present for Harry's annual cystoscopy, performed at Mount St. Mary's Hospital in Lewiston. There was no reason to expect a problem; Harry had been clean for fifteen years. Still, he seemed nervous as we sat in the waiting room. He was wearing khaki cargo shorts, a green-and-blue striped polo shirt, and white New Balance sneakers. Just before 11 a.m. on Friday, September 20, the nurses called him back to get prepped.

Diane and I stayed in the waiting room. She talked about the loss of her three younger siblings in the 1960s. Mark was feverish and "covered in hives" the night he died of spinal meningitis. Relatives who came to the hospital to pay their respects were weeping; Diane wondered why everyone was so sad, given that she'd been assured Mark had gone to be with Jesus. The conversation was interrupted by a nurse, who said we could see Harry before he went under. He was in a hospital gown, lying on a gurney. "I hope that tumor didn't come back," he said groggily. The nurse asked Harry his name. "Herbie White," he joked. Diane and I returned to the waiting room, and she picked up her stories. Dona was a "tiny little thing with dark hair" and a "gaping sore" on her back. "When she was born, she didn't have an open anus; they had to open it [surgically]," Diane said. "But her heart was really strong." Diane had nightmares in which one of the children—sometimes Mark, sometimes

John or Dona—was climbing out of a casket. "I was just a kid," she said. "Three deaths in a row was a lot. My mom said if she didn't have her faith, she doesn't know how she would have lived through it." Dottie's doctor suggested she not have any more children because of the chemicals Ray worked with. She disregarded this advice, and Ray Jr. was born on October 21, 1966. "How long is God going to let us keep him?" Diane asked her mother when the baby came home.

Around 12:30 p.m. Diane and I were summoned to recovery. Harry was still groggy. His urologist, Brian Rambarran, told Diane he'd found an eight-millimeter tumor on Harry's bladder and excised it. Diane relayed the news to Harry, whose eyes widened. "What? Son of a bitch." Diane rubbed Harry's neck as he slid from the gurney into a wheelchair and received his discharge instructions from the nurse. He understood the upshot of the doctor's discovery: he'd need a cystoscopy every three months for at least two years. The anxiety would come back. Rambarran, full of nervous energy and wearing a "Rambo" vest over his scrubs, said he'd sent a slice of the growth to the pathology lab. Maybe it's benign, Diane said. Harry wasn't buying it. "The last time I had a tumor it was malignant," he said glumly. Four days later, the pathology report came in. "HIGH-GRADE PAPILLARY UROTHELIAL CARCINOMA," it read. "High-grade" meant the tumor could grow quickly and should be treated aggressively. Rambarran tried to convince Harry to stay in New York a while longer to begin chemo, but Harry declined and returned to Las Vegas with Diane.

The years 2020 and 2021 were merciless for the Kline and Weist families. Dottie broke her hip, then her femur. Both required surgery; it took her two days to emerge from the anesthetic fog after the latter, a period during which she was doing bizarre things, such as reflexively removing her hospital gown. When she came home, her blood sugar was low; she fainted and broke her right ankle. Fearful of falling out of bed, she began sleeping in a recliner in her living room. Ray, who slept on the sofa beside her, had problems of his own. In early 2021, he had trouble urinating, had to be catheterized, and underwent surgery to clear a blockage caused by scar tissue in his ureter. For several months his urine was routed into a bag. By year's end blood was coming out of his penis. He was in the worst pain of his life—so intense he uncharacteristically complained about it. A cystoscopy—performed, as always, without anesthesia—detected what Dottie called "a fairly large tumor. It was pumping blood out like crazy. The doctor cut that out and cauterized it." Ray still couldn't urinate, and his scarred ureter wouldn't allow the insertion of a Foley catheter. His bladder was nearly at capacity, causing him more agony. A doctor who specialized in catheterization was summoned and finally was able to insert the tube. "Ray thought the guy was going to break his back, trying to get that Foley in," Dottie said.

Harry had three cystoscopies in 2020. The first two came up clean. The third found two high-grade tumors, which Rambarran removed. A cysto in April 2021 detected no growths, but Rambarran decided to take a biopsy of Harry's prostate because his prostate-specific antigen, or PSA, numbers were up. The biopsy showed cancer. Harry met in May of that year with an oncologist who gave him three options: take hormone therapy, which would arrest the tumor by halting production of testosterone; have his prostate removed; or do nothing and hope the cancer didn't progress. Harry rejected the first two because they'd likely cause erectile dysfunction and, as he liked to put it, "I still have some gas in the tank." He chose option three, agreeing to undergo regular testing. He and Diane went home to Las Vegas, returning to New York in July so Harry could have another cysto. That procedure took place July 12, and there was no evidence of a bladder-cancer recurrence. Later that month, however, both he and Diane came down with COVID-19, though they'd been vaccinated. Harry texted me on July 28: "At the hospital Dianes not doing good. I'm very worried." Diane's fever had spiked to nearly 105 degrees, and she was weak, with low blood sugar. Harry had taken her to Mount St. Mary's; she stayed about thirty minutes and was released after her temperature dropped below 100 and her vital signs were deemed normal. That evening, Harry had driven Diane by her parents' home, a few minutes from the hospital. She rolled down the window, blew them a kiss, and said she loved them.

Harry called me mid-morning on the 29th to say that Diane had died around 4:30 a.m. He could hardly speak. The two of them had been quarantining in Ray Kline Jr.'s twenty-four-foot RV, parked outside his house in Newfane, thirty-five minutes northeast of Lewiston. Diane had gotten out of bed to go to the bathroom, taken one step, and fallen. Harry suspected she was dead before he left to get his brother-in-law. When the two men returned to the RV, they found Diane motionless, wedged between the bed and the wall. Her teeth were clenched. Her face was blue. Ray Jr., a banking technology specialist about eight years younger than Diane, performed CPR on her for fifteen minutes while waiting for the EMTs. When they arrived, there was nothing to be done. Ray Jr. called his mother to tell her what had happened and heard her wail, "Not the fourth!" Ray and Dottie had lost their fourth child. Harry had lost his partner of almost forty-two years.

Harry spent the rest of 2021 trying to navigate life without Diane, who managed the couple's finances, remembered all the family birthdays, and prepared lavish holiday dinners. He felt guilty. "I feel if I wouldn't have went home for those test[s] Diane would still be here," he texted me on August 22. There was a "crying" emoji after the message. Harry held a memorial service for Diane on September 13 at the Hardison Funeral Home in Ransomville, New York. A spillover crowd was welcomed by Pastor Kevin Robbins of Lockport Alliance Church, and the rituals of reminiscence—funny stories, sudden breakdowns—began. Harry and Diane's eldest daughter, Holly, lost her composure as she read something from a friend in Colorado, as did Kristan, the youngest, following a crack about her being a "union girl" who took after her father. There were many references to Diane's strong faith and her ability to make anyone—even gawky teenagers—comfortable from the moment she met them. One of those former teenagers made a joke about the Weists' hard-to-close bathroom door, which drew knowing laughs.

Then it was Dottie's turn. Ray Jr. helped her to her feet and she turned around to face the group, leaning on a cane. She spoke evenly. While others were remembering Diane's mortal life, "I'm celebrating her eternal life," Dottie said. "I know where she's at." After two more family friends spoke, Harry stood up. He'd been dabbing at his eyes with a handkerchief off and on during the service. "Let's see if I can get through this," he said. He recounted his and Diane's friendship at Goodyear and their early days as a couple. When they decided to wed only seven months after they began dating, many predicted the marriage wouldn't last more than a couple of months. He broke down as he said it had lasted more than four decades. His father-in-law sat expressionless during the entire service. He did not speak, and no one who knew him expected him to.

Harry weathered the fall of 2021 by leaning on his children, the youngest of whom, Kristan, was living with him as she and her husband, Tony Nigro, had a home built in Henderson, Nevada, a Las Vegas suburb. This arrangement gave Harry access to his six-year-old grandson, Tony Jr., with whom he was close, and kept him from lapsing into paralyzing desolation. Still, there were moments. He and Diane had been inseparable. It was she who insisted that Harry keep getting cystos for his bladder cancer, often over his protests. She was calm but forceful, and Harry listened to her. She was the one who remembered details—someone's name, when something happened.

Harry also found succor from the Kline family. He was like a son to Ray and Dottie, an older brother to Ray Jr., who, by his own admission, was "a little bit of a mama's boy." Harry, not physically imposing but fearless in his youth, had taught Ray Jr. how to defend himself. This knowledge came in handy at the Walmore Inn, a tavern in Sanborn where Harry and Ray Jr. hung out, and at Rich Stadium. Both venues were breeding grounds for alcohol-fueled fights. Ray Sr. remained a reflection of his upbringing in Pennsylvania's coal country, where workers were paid by the ton, and if there was no work on a given day, there was no pay. He was grateful to Goodyear, despite what it had done to his body and, perhaps, two of his five offspring. "He was reluctant to bad-mouth what transformed our family," his son said.

Harry felt no such loyalty. In October of 2021, three months after Diane died, Goodyear informed him his monthly health insurance premium would go up $100, to $700, even with Diane off the policy. "They're taking a third of my pension now," Harry complained. A cysto in November came up clean, but a few weeks later Harry was passing blood when he urinated. Rambarran assured him this was normal following the procedure. Ten days before Christmas, Harry awoke to an empty bed at 3 a.m. and for a moment found himself in a dreamlike state, thinking Diane was still alive. *She's probably sleeping on the couch downstairs,* he told himself. Reality intervened, and Harry remembered Diane was gone. "That really put me in a tailspin," he said. "I can't believe how fast life turned. I knew Diane was not going to live to a ripe old age but I didn't think she would go at sixty-two. My brain digs up all these memories. The file comes open and all this stuff comes out."

CHAPTER 22

WODKA FINISHES
THE FIGHT

TEVE WODKA ENTERED the final month of 2021 with four bladder-cancer cases pending against DuPont—those of Jim Sarkees, Doug Moss, Robert Mooney, and Gary Casten, the last of whom had died. Sarkees and Moss hadn't had any recurrences of their tumors, but two kinds of cancer had become embedded in the surface of Mooney's bladder, and he'd suffered a stroke. In July, H. Kenneth Schroeder, the federal magistrate judge in Buffalo, had finally ruled in the Moss case after almost three years, and the outcome was not good for the plaintiff. In considering DuPont's motion to dismiss, Schroeder held that a jury could hear liability evidence against the company only through 1990. This meant that DuPont would be off the hook for Goodyear workers first exposed to ortho-toluidine from 1991 on, even though DuPont—right up to the time it stopped making the chemical in 1995—continued to tell Goodyear that the workplace air concentration limit of 5 ppm was safe. DuPont failed to mention that its own operating procedures forbade *any* ortho-toluidine in a worker's urine. Wodka appealed Schroeder's ruling to US district court.

He had been sparring with DuPont all year, trying to pry loose internal documents he believed would show the company's state of mind between 1993, when its own calculations showed exposure to ortho-toluidine at the legal limit would produce levels of the chemical in urine thirty-seven times higher than what NIOSH had detected three years earlier, and 1995, when DuPont assured the EPA that there was no need to adjust the OSHA exposure limit. By November, Wodka let it be known that he was prepared to file a motion to compel DuPont to hand over the documents. Soon after, the company's lawyers broached the possibility of settling all four cases, and by mid-December the parties had reached an agreement in principle. There were another six weeks or so of haggling over details, and Wodka was waiting for the deal to collapse. Finally, on February 4, 2022, it was consummated, the details confidential as always.

Jim Sarkees and his wife, Debbie, spoke with me four days later from their home in Charleston, South Carolina. Jim had retired after forty years as a truck driver. He'd grown up in Niagara Falls, graduated from high school in 1973, gotten married, and enrolled in a two-year technical college. He left after a year to make a living. His father, Haleem, who worked in Department 245 at Goodyear, got him a summer job in 1974. There he was introduced to the monotony of Nailax bagging and the repugnance of cleaning the reactors and Sparkler filters. The filter cleaning was performed in a breezeway because the orange vapors were so overpowering. Sarkees had no respiratory protection and received no information about the constituents of the fumes. He'd have to turn away to catch his breath.

Cleaning the muck from the reactors was equally bad. Sarkees, lithe and uncomplaining, would be tapped to enter the vessels after they had been cooled to chisel out the solids. Wearing coveralls and a cartridge-style respirator that covered his nose and mouth, he would enter a manway and descend a wooden ladder to the bottom, as would a coworker. They'd put the loosened material into a pail that had been lowered on a rope; after a while the fumes would force them outside, to Fifty-Sixth Street, for fresh air. Sarkees would hear intermittent warnings from his coworkers: make sure your fingernails don't turn blue. It meant nothing to him at the time.

Sarkees was asked to stay on at Goodyear at the end of the summer because the lab was shorthanded. He found lab work less objectionable than production; it entailed testing product samples from the tank farm for purity and didn't make people's extremities change colors. He held this job the rest of the year, then left Goodyear for good. All told, he spent seven months at the plant.

Sarkees's next encounter with Goodyear came in 1998, when he was notified by letter that he qualified for urine screening, given the high bladder-cancer risk for those who had worked in Department 245. He was inclined to dismiss the invitation, figuring he'd been at the plant for such a brief period that he was in no danger. But his second wife, Debbie, told him he had nothing to lose by participating. He gave the sample and returned it to Goodyear in the FedEx envelope the company had provided.

Sarkees repeated the process every six months; nothing of consequence happened until November 2003, when he learned that abnormal cells had been found in his urine. He underwent his first cystoscopy at Roswell Park in Buffalo and came up clean. More suspicious cells were detected in May 2012; again, a cysto showed nothing. The cells reappeared in November 2014, and this time Sarkees chose not to have a cysto. "It was like a roller coaster," Debbie said. "They'd find atypical cells; we'd go to Roswell and they wouldn't find anything. Then the big one hit."

The "big one" came in October 2016. Sarkees's urine test through Goodyear had found another anomaly, unconfirmed at Roswell Park. Khurshid Guru, the cancer center's urology department chair, nonetheless urged Sarkees to have a cysto—his first in more than four years—as a precaution. It unearthed a high-grade tumor, which the doctor removed. Sarkees began Bacillus Calmette-Guerin, or BCG, immunotherapy, which involves the injection of a solution infused with the bacterium used in the tuberculosis vaccine directly into the bladder through a catheter. The liquid is held there for an uncomfortable two hours while the immune system, roused by the germ, attacks the cancer cells. BCG treatments made Ray Kline very sick and Jim Sarkees very tired. Sarkees had fifteen of them, along with fourteen more cystos. All but one came up clean, and the frequency of the tests diminished. By July 2021, Sarkees was back to having one cysto a year, a reward for being tumor-free. Debbie was not reassured. "It's terrifying," she said. "It's so long between [cystos] now that I'm afraid [the cancer] is going to get a jump on him before they see it. They've always talked about not *if* it comes back, but *when* it comes back." Said her husband: "It kind of feels like you're being chased."

Gary Casten was overtaken before he could see the fruits of his lawsuit. Casten started in Department 245 in 1965 and retired in 2004. He was a Nailax bagger and reactor operator and, as such, was exposed to "some nasty stuff," he recalled. "The old joke was, 'Don't worry, nothing here will hurt you.' You just worked with it and hoped it didn't." In 1970, Casten's fingertips turned blue during a shift; he was diagnosed with aniline poisoning and hospitalized. Otherwise, he escaped serious illness or injury. He watched his friends and coworkers Hank Schiro and Charlie Carson die of bladder cancer, had post-retirement bouts of skin and prostate cancer and, after seeing Rambarran for what he thought were kidney stones in August of 2020, learned that a large tumor was lodged in the renal pelvis of his left kidney. It was essentially the same cancer that had struck the other Goodyear workers, the difference being that Casten's originated at the top of the urothelial tract instead of the bottom. He had a kidney removed in September, underwent chemo, got the all-clear from an oncologist in November and sounded upbeat when he spoke to me shortly thereafter. By the time he gave a deposition in April of 2021, however, he was feeling lethargic and shedding weight. The tumor had begun to metastasize, and he was back on chemo. During the deposition, Wodka asked Casten if he knew what would come next. "I never asked how long I have or what the prognosis is," he replied. "I'm really not ready to approach that question yet." Three and a half months later he was dead.

Sixteen days before Donald Trump left office in January 2021, Wodka filed a complaint with the EPA on Casten's behalf. The complaint alleged that DuPont had been "in possession of substantial risk information" about ortho-toluidine since April 20, 1993, the date industrial hygienist Thomas Nelson reported internally that airborne exposure to the chemical at the legal limit resulted in urinary concentrations far above what NIOSH had found in the Goodyear workers. DuPont "knowingly misinformed" the EPA that the OSHA standard was protective not quite two years later, Wodka wrote, and had "willfully violated" section

8(e) of the Toxic Substances Control Act since April of 2018, when it was put on notice by a filing in the Moss case. The section requires that a manufacturer report "the discovery of previously unknown and significant exposure to a chemical" when such exposure is "combined with knowledge that the subject chemical is recognized or suspected as being capable of causing serious adverse health effects (e.g., cancer, birth defects, neurotoxicity)." Wodka concluded his letter by noting that "tens of millions of pounds" of ortho-toluidine were still being imported by the United States each year. "DuPont's own research proves that legal exposure at 5 ppm is a lethal level of exposure," he wrote. "Government action to protect workers should have occurred a long time ago. DuPont's failure to make an adequate and timely reporting under Section 8(e) has contributed to this delay." Had DuPont told the EPA about Nelson's discovery in 1993, as Wodka maintained it should have, the agency had procedures in place to relay the information to OSHA and NIOSH. The outcome might have been a stricter airborne limit for the chemical at Goodyear, which had blithely assumed the very low readings it was getting in Niagara Falls meant its workers were safe. It was a blunder that may have cost lives.

DuPont had been stung for a similar transgression in 2005, agreeing to pay an EPA civil penalty of $10.25 million—at that time a record sum—for failing to report health risks posed by perfluorooctanoic acid, or PFOA, discharged from its Washington Works in West Virginia. Wodka saw nothing less despicable in DuPont's handling of the data its own employee had unearthed about ortho-toluidine. "There is a direct connection between DuPont's failure to abide by the [Toxic Substances Control Act] and the continuing cases of bladder cancer in the Goodyear workers in Niagara Falls, New York," Wodka wrote in his letter. There was no response.

In early March of 2022, Wodka sent an email to the EPA's assistant administrator for chemical safety and pollution prevention, Michal Freedhoff, asking if there was any record of his complaint. A staff member replied that there wasn't, suggesting the Trump people hadn't logged or otherwise noted it. The same month, Wodka and three other veterans of the bladder-cancer wars filed a petition with OSHA, asking that it significantly tighten the ortho-toluidine exposure limit of 5 ppm. Science had shown this number, adopted by OSHA when the agency came into being in 1971, to be dangerously out-of-date, the petition said. Leaving it in place was tantamount to endorsing "a lethal level of exposure." Austria and Switzerland, by way of contrast, had limits of 0.1 ppm—fifty times stricter. OSHA didn't respond to the petition.

In late March, Wodka got the acknowledgment he'd been seeking from the EPA: Gloria Odusote, representing the agency's Waste and Chemical Enforcement Division, thanked him for reporting DuPont's alleged 8(e) violation. "We are looking into it," she wrote in an email. Six months passed. Finally, in late September, Wodka heard again from Odusote, who wrote, "We did not take further enforcement action [against DuPont] because we had a document that demonstrated they met their 8(e) obligations." The details, she explained, were considered "confidential business information." Wodka filed a Freedom of Information Act request at the end of 2022, asking for a copy of the document. This earned him a thirty-five-minute video call with Odusote and three other EPA staff members in January 2023. After the call ended, Wodka was disappointed. The Toxic Substances Control Act, requiring manufacturers to divulge unpublished scientific data identifying substantial risks to human health, had been watered down in successive interpretations by the EPA since its passage in 1976. The only exception to the law was if the same information had been published in the open scientific literature or had already been reported to the EPA by someone else—an objective test. To Wodka, it was indisputable that Nelson's unpublished 1993 calculation should have been reported. But the EPA hadn't gone after DuPont for a violation, he was told, because one of the agency's toxicologists had been able to replicate Nelson's calculation in 2022. Wodka found this explanation indefensibly subjective, clearly counter to what the law required. The EPA's response to his FOIA request in February 2023 didn't make him feel any better. The documents he received showed that data from two DuPont studies underpinning Nelson's work were submitted to the EPA as confidential—improperly, in Wodka's opinion. This kept crucial scientific information from the government, industry, and the public.

Wodka's belief that ortho-toluidine remains a menace at Goodyear in Niagara Falls was affirmed by a NIOSH study published in the *American Journal of Industrial Medicine* in May 2021. The researchers' findings were jarring: even after all the improvements Goodyear had made, resulting in airborne levels of the chemical that averaged 11.2 parts per *billion*—a tiny fraction of what OSHA allowed—plant workers faced a lifetime bladder-cancer risk twelve to sixty-eight times higher than what the Supreme Court had deemed legally acceptable in a landmark 1980 case involving OSHA's benzene standard.

The NIOSH paper motivated Wodka, who had just turned seventy-two, to offer to help the United Steelworkers prepare a grievance against Goodyear over continuing ortho-toluidine exposures. He saw it as one of the most important things he could do for the Goodyear workers at this stage in his career. The offer was accepted, though Wodka's formal relationship with the union had ended in 2015. His position was that the ongoing risk identified by NIOSH represented a violation of the collective bargaining agreement, under which Goodyear was expected to make "reasonable provisions" for the health and safety of its employees. Correspondence between the parties began, and for a time there was reason for hope. In a letter to Local 4–277 president Joe White and Wodka dated July 13, 2021, Mark Kitchen, the plant's human resource manager, wrote that "Goodyear is committed to ongoing reduction of ortho-toluidine exposure at [its] Niagara Falls facility." The company would bring in a "global industrial hygiene provider" to provide a "comprehensive assessment" of the operation, Kitchen promised. White and Wodka were skeptical but agreed to reserve judgment until they saw the report.

Their skepticism was well founded. On August 31, two representatives of Chubb Global Risk Advisors, affiliated with Goodyear's insurer, performed what they called a "limited on-site qualitative assessment" of the plant. In their report, dated October 14, they recommended that Goodyear enforce the wearing of personal protective equipment, keep plant surfaces clear of ortho-toluidine, and make sure ventilation systems were working properly. The hygienists did no air sampling. White and Wodka expressed their dissatisfaction with the exercise in a letter to Kitchen five days later. They'd asked that an independent consultant be brought in to find ways to separate the workers from the chemical that had stalked them for more than sixty years; DuPont had created this separation with the installation of its Chambers Works glove boxes in the 1960s. What the Chubb men delivered was a half-baked, company-friendly statement of the obvious. To Wodka, the risk was clear: if conditions stayed as they were, more workers would likely get sick. At that point he was cut out of the grievance process, perhaps because he was pressing the union to do something it was uncomfortable doing. Wodka wondered if leaders of the Steelworkers had given in to the irrational fear that Goodyear would close the plant, even though it had made no such threat.

In fact, the Steelworkers did file the grievance, but put it on hold to give the company time to make changes in plant operations, such as replacing inefficient ventilation systems. Paring ortho-toluidine exposures to "as low as reasonably achievable" and applying the hierarchy of controls would be a lengthy process, the Steelworkers' director of health, safety, and environment, Steve Sallman, told me. "This is going to be a marathon, not a sprint." Wodka doubted that the union would push Goodyear as far as it needed to be pushed—to ensure there was no worker exposure, period.

To former OCAW president Bob Wages, all of this seemed familiar. A contemporary of Wodka's, Wages joined the union in 1969, when, as a twenty-year-old college student, he went to work at the Phillips oil refinery in Kansas City, Kansas. He was part of a wave of younger, more vocal workers who went into the business as their aging predecessors were retiring or dying off. Conditions were still abominable; Wages and his coworkers, for example, were told to clean asphalt tanks with fuel oil, a ludicrously unsafe process that made them dizzy within minutes. He complained to his union reps, to no avail. Nor did he get anywhere when he complained about having to tear out asbestos-laden refractory insulation wearing only a 3M paper mask. In the spring of 1972, Wages, then only twenty-three, ran for president of OCAW Local 5–604 and won. He graduated from the University of Kansas in May and started law school at the University of Missouri–Kansas City that summer. He kept working at Phillips until he got his law degree and passed the bar in 1975. A year later, he joined the legal staff at OCAW headquarters in Denver and almost immediately was thrust into the Pittsburgh Corning asbestos fiasco in Tyler. This is how he came to know and admire Mazzocchi and Wodka. He would work with them on other imbroglios: the forced sterilization of lead-exposed female workers at the American Cyanamid plant in West Virginia, and the accidental sterilization of male workers exposed to the pesticide dibromochloropropane—DBCP—at plants run by Occidental and Dow Chemical.

Wages rose through the ranks at the union, becoming vice president in 1983 and president in 1991. He saw it merge with the United Paperworkers International Union and, after he retired to practice law in 2001, be absorbed by the Steelworkers. Both adjustments weakened the union, in his view. He'd followed the bladder-cancer situation at Goodyear's Niagara Falls plant while leading the OCAW but hadn't kept up with it. When I told him what had happened with the Steelworkers' grievance, he was suspicious. "The idea that you file a grievance and hold it in abeyance is just bullshit," he said. "That smacks of 'cover your ass.'" He'd heard many times during his OCAW tenure that a particular plant might close if the union pressed too hard for controls on toxic exposures. "Our response every single time was that we can't let people die," Wages said. "If the tradeoff is for people to die, then maybe the plant has to go."

CHAPTER 23

WODKA'S CRUCIBLE

THE MONTH AFTER he settled his last four cases against DuPont, Steve Wodka was admitted to NYU Langone, a highly regarded hospital on Manhattan's East Side, for his third heart procedure in nineteen years. Wodka had been born with mitral-valve prolapse—a heart murmur—that caused blood to leak from one chamber of the heart to another. He'd had the mitral valve repaired in 2003 and in 2014 underwent aortic-valve replacement; I'd visited him at his home in New Jersey not long before the latter procedure, and he seemed uneasy. If he died on the operating table, he joked, I should tell his wife I had his permission to dig through the 170 boxes of papers in his basement. He survived, of course, but the murmur persisted, and by 2021 it was clear the repaired mitral valve needed to be replaced. This sent Wodka back to NYU Langone a third time. What happened there gave him a taste of what some of his Goodyear clients had endured.

Wodka entered the hospital on Friday, March 25, 2022, and underwent eight hours of surgery. When he came to, the tube that had allowed him to breathe during the operation was lodged in his throat. After the first two surgeries, Wodka said, the tube had been removed within minutes. This time it stayed in place for six hours, giving him the sensation of drowning. "I would scream as hard as I could. I'd thrash around and say, 'Take it out,' and they would have the nurses sit on top of me." The explanation for this unpleasantness came later: his lungs weren't functioning properly. Wodka's heart rate fluctuated wildly over the weekend, and by Monday the 28th a team of doctors had concluded that he needed a pacemaker. That day, Wodka passed out twice in his hospital bed. He awoke after the first blackout to find his room filled with doctors and nurses. One of the doctors shouted orders, in a scene reminiscent of overwrought TV medical dramas. The second time he passed out, "I thought, 'This is it.'" His heart stopped beating for eleven seconds.

Even with the pacemaker installed, Wodka didn't feel right. He was weak, and his heart rate was stuck at 115 beats per minute—way too high. A doctor in New Jersey performed a cardioversion—which uses low-energy electrical shocks to restore a regular heart rhythm—and only then did Wodka stabilize. His ordeal, he said, had brought to life the medical reports he'd read on men like Hank Schiro, who'd had his bladder removed in 1980 after his cancer kept coming back. Even this disfiguring procedure, known as a radical cystectomy, couldn't save Schiro; the metastatic process was under way, and he would die six years later.

Wodka had recently turned seventy-three and all but decided to retire at the end of the year. He'd begun digitizing the documents in his basement and already stumbled onto something surprising: Schiro, it turned out, hadn't been the first Goodyear worker in Niagara Falls to be diagnosed with bladder cancer, in 1972, as NIOSH incorrectly stated in its first interim report in 1989. That distinction belonged to a man named Edward Babut, who'd started working in Department 245 in 1963 and was diagnosed in 1973, a year before Schiro's cancer was actually found. Records from Niagara Falls Memorial Medical Center show that Dr. Joseph D'Errico performed a cystoscopy on Babut on June 14, 1973, and found a "very well-defined unmistakable tumor in the mid-left bladder floor." D'Errico had ordered the test after a urine sample Babut gave as part of his annual Goodyear physical showed the presence of abnormal cells. The pathology report on June 20 confirmed the presence of a Grade II tumor with submucosal invasion, meaning the malignancy was moderately aggressive and had migrated to the layer of connective tissue beneath the mucous membrane.

At the doctor's suggestion, Babut filed a workers' compensation claim against Goodyear. But he never hired a lawyer, D'Errico never put his suspicions about the work-relatedness of Babut's illness into writing, and the case was closed on November 20, 1973, with the notation that there was "no medical evidence" to support it. Would the narrative have changed if the claim had gone forward? Not necessarily. In 1973, NIOSH was still fifteen years away from making its first visit to the Niagara Falls plant. Ortho-toluidine hadn't been flagged by the government as a bladder carcinogen; there was only animal evidence and the yet-to-be-translated Soviet study from 1970. From Wodka's perspective, however, there *was* enough evidence in 1973 for DuPont to have issued a warning about the chemical to its customers. That wouldn't happen for another four years, and when it did, it wasn't done with any urgency or honesty about the science. Wodka was retained by Babut two months before his death in 1988 and years later negotiated a payout from DuPont and the other ortho-toluidine suppliers, which went to Babut's younger brother, Henry.

I found it hard to believe that Wodka really would retire, given his decades-long obsession with the DuPont cases, but he seemed serious about it. He'd won compensation for dozens of wronged Goodyear workers and their families and was especially proud of a class-action lawsuit he'd filed against the company in 1993 and settled five years later. The settlement required Goodyear to semi-annually test some five hundred former workers and retirees—the size of the class at the time—for early signs of bladder cancer. The company had been screening employees for the disease but, in response to an OCAW proposal in 1990, said it was "not interested" in extending the program to those who'd left the Niagara Falls plant.

Wodka kept up the pressure, noting that retirees were at particularly grave risk because they'd received the highest exposures to ortho-toluidine and experienced the longest latency periods. Goodyear ultimately abandoned its position and committed to testing, for the rest of their lives, all Niagara Falls workers after they retired, quit, or were fired.

Wodka remains the court-appointed plaintiffs' class adviser for the Goodyear plant in Niagara Falls and the former Morton International—now Dow—plant in Paterson, New Jersey, where Joe Nicastro had worked. In July 2022, he was notified that the Goodyear screening program had found another bladder-cancer case, diagnosed in late 2021. Wodka didn't know the victim's name; he knew only that it was someone who had been exposed to ortho-toluidine for at least a year prior to June 11, 1990. By his count, this was case No. 78. Twenty-two of those cases were caught early as a result of the expanded Goodyear screening program.

While DuPont was his target in the courts, Wodka maintains that Goodyear could have done more to stop the onslaught of cancer. "They knew more at the critical times than they let on," he said. "Had they been proactive, had they been honest about what they knew, you would think that most of the cancer cases could have been avoided." I asked spokespeople for both companies to weigh in on the Niagara Falls episode; each gave me a written statement.

DuPont's was brief. "As you may know from your review of litigation," Daniel Turner, the company's director of public affairs, wrote in an email, "the cases involve only E.I. du Pont de Nemours & Co., not DuPont de Nemours, which is a separate company that was established in 2019. As a result, we will respectfully decline comment." Turner was referring to the breakup of the legacy company, which merged with Dow in 2017, into three pieces: Dow, DuPont de Nemours, and Corteva. Before the Dow merger, DuPont had spun off its chemical division and called it Chemours. DuPont's then CEO, Ellen Kullman, assured shareholders this would spawn a "leaner and more efficient" DuPont that would "help solve major global challenges." But many, including Wodka, believe Chemours was created to absorb DuPont's liabilities for misdeeds such as dumping per- and polyfluoroalkyl substances—PFAS— into waterways. "By shunting the chemical factories into an independent company," the *New York Times* reported in 2021, "DuPont would be insulated from future lawsuits related to those chemicals."

Goodyear's statement to me was more expansive. "Goodyear remains committed to actions to address ortho-toluidine exposure inside our Niagara Falls facility," Connie Deibel, the company's communications manager for global operations, wrote. She continued:

> We have worked closely and transparently with our associates, the local union and the National Institute for Occupational Safety and Health (NIOSH) to better understand and manage potential exposure.

When Goodyear learned of the concerns associated with the use of ortho-toluidine, we adjusted our systems and processes and put engineering controls in place in the 1980s to further reduce levels in the plant. As new information and processes become available, Goodyear—working in full cooperation with our associates, the union and appropriate agencies—has made and will continue to make improvements to protect our associates. The biannual bladder cancer screening available to all active, previously employed, and retired associates at the Niagara Falls facility has been ongoing for more than two decades—at no cost to our associates or retirees. Although not required by any regulatory agency, we also worked closely with the local union to develop a pre- and post-shift urine screening process to determine contribution of the chemical agent through skin contact with complete transparency for every group involved. While we believe our safety and industrial hygiene procedures—including the required use of engineering controls and personal protective equipment—are working, we will continue to work closely with NIOSH and our local union.

CHAPTER 24

WORKERS ARE
(MOSTLY) ON THEIR OWN

THINGS MIGHT LOOK LESS BLEAK in American workplaces if OSHA could do its job the way the law envisioned, setting and resetting chemical limits as the science dictates, enforcing limits already on the books, using the act's catchall general duty clause—which ensures the right to a workplace "free from recognized hazards"—when specific standards don't exist. Among OSHA's leaders, Eula Bingham probably came closest. But, through no fault of her own, her 1978 benzene standard was stayed by the Fifth Circuit Court of Appeals after a challenge by the American Petroleum Institute, and later resulted in a Supreme Court decision that officially assigned workers second-tier status with respect to health protections. OSHA, the upper court held in 1980, had to show that "a significant risk of material health impairment" existed before the agency could issue or adjust a chemical standard. "Significant" turned out to be one extra case of cancer per thousand workers. A chemical found in the general environment, by way of contrast, was considered worthy of control when the excess cancer risk was one in a million. The Fifth Circuit stay and the Supreme Court's decision to overturn the benzene regulation kept the 1 ppm limit from taking effect until 1987, when OSHA issued an amended standard. During the nine years that the limit was stuck at 10 ppm, two researchers later estimated, between 30 and 490 workers were exposed to enough benzene to die of leukemia.

More would likely die from lymphoma or aplastic anemia, diseases also linked to the chemical.

In January 1989, OSHA set or tightened 376 chemical exposure limits in a single rule, a bold move the agency estimated would eliminate fifty-five thousand work-related diseases and 683 deaths each year, at an annual cost of $150 per worker and $6,000 per affected plant—"a fraction of 1% of the sales for all affected industry sectors." Legal challenges by both industry and labor—the former thought some of the exposure limits too strict, the latter too lenient—led to a devastating loss three years later: the Eleventh Circuit Court of Appeals vacated all the new limits, saying OSHA had failed to demonstrate that significant risks existed under the old ones. Reagan appointee John Pendergrass was behind the attempt to circumvent the ineffective, one-chemical-at-a-time method of regulation used to that point, and little of consequence has happened since. In 2010, thinking it might make headway under President Barack Obama, OSHA sought online comments from the public on which substances were most in need of revised or first-ever standards. Certain ones were mentioned multiple times: carbon monoxide, chemotherapy drugs, diesel engine exhaust, manganese, lead. Nothing came of the exercise. OSHA did push through an updated standard for silica— the bane of miners, stonecutters, and sandblasters—before the end of Obama's presidency. It issued a beryllium standard just before Donald Trump took office in 2017, but only because Materion Brush, the world's biggest manufacturer of the metal, and the United Steelworkers, whose members were at high risk of exposure, signed off on a deal beforehand. Otherwise, OSHA has all but thrown up its hands. In an extraordinary 2013 press release, it suggested employers consider adhering, voluntarily, to exposure limits recommended by NIOSH or set by the state of California rather than rely on all-but-meaningless federal numbers for compounds like ortho-toluidine. This assumes there *are* numbers; in most cases there aren't. There are only 470 workplace-exposure limits, most of which were simply adopted by OSHA when it came into being in 1971 and are based on old, and sometimes very old, science. The agency has issued only thirty-eight major health standards—less than one a year—since its creation.

OSHA suffered mightily at the hands of Donald Trump. By the end of his administration, the number of inspectors had plunged to nearly its lowest level since the agency was created. There were only 1,719—755 federal, 964 state—to cover 10.4 million workplaces in fiscal year 2021, according to the AFL-CIO. This translated to one inspector for every 81,427 workers. OSHA's puny budget meant the nation was spending $4.37 to protect each worker—a little less than the cost of a venti latte at Starbucks. As of the end of 2022, no new chemical standards were being proposed. The number of discrete substances regulated since 1970 stood at eighteen, not counting the 1974 rule covering fourteen carcinogens. "OSHA's work remains stalled on chemicals," the AFL-CIO said charitably. It said this as the science on some familiar poisons is looking increasingly ominous. The workplace standard for benzene, for instance, is 1 part per million over an eight-hour workday. That was a bold number when Eula Bingham put it forward. In 2021, however, a highly regarded scientific organization, the American Conference of Governmental Industrial Hygienists, said the eight-hour limit should be 0.02 ppm—*fifty times* stricter than what's permissible today. Its rationale: benzene still threatened an estimated 230,000 workers in the United States and had shown "toxic effects on the bone marrow at exposures lower than previously considered relevant." In 2018, the European Chemicals Agency recommended a limit twenty times stricter than what OSHA allows. Will OSHA do anything? The odds are nil, or close to it.

Amid the regulatory paralysis, new workplace threats are emerging. Among them are nanomaterials—particles or fibers one ten-thousandth the width of a human hair. These industrial wonders can be found in baseball bats, bicycles, cosmetics, intravenous drugs, dust-repellent coatings on windows, and antifungal coatings on sports socks. Miniature cylinders of carbon atoms known as carbon nanotubes are so light and strong that they're used in airplane wings, yacht masts, and bulletproof vests. But some nanotubes behave like asbestos when injected into mice and cause fibrosis and mesothelioma. Will they do the same in humans? Should we wait a few decades to find out? NIOSH has recommended an exposure limit for carbon nanotubes of 1 microgram per cubic meter of air over an eight-hour workday, but there isn't an enforceable OSHA standard. From 2012 to 2014, NIOSH researchers surveyed twelve nanomaterial-manufacturing facilities with a collective 108 employees and took 240 air samples. Ten percent of the samples were above the recommended limit. No acute effects were noted among the workers, but the investigators weren't looking for fibrotic lung disease or other evidence of chronic illness. (The agency is trying to close this knowledge gap by compiling a registry of workers exposed to carbon nanotubes in hopes of assembling a cohort for an epidemiological study.)

The COVID-19 pandemic kept NIOSH from conducting further field studies. It did, however, come up with a recommended exposure limit for silver nanomaterials—known for their antimicrobial capabilities and used in soaps, paints, cooking utensils, medical instruments, and many other products—of 0.9 micrograms per cubic meter. It acted because the substances produced lung inflammation and liver bile duct hyperplasia (increased cell production, a possible precursor to cancer) in rats. Nanomaterials do plenty of good—they were what made the Pfizer-BioNTech and Moderna COVID-19 vaccines so effective. But their capacity for causing biological mischief in humans is not well understood.

Compared to the United States, the European Union has moved to bring chemicals to heel at a headlong pace. In April 2022, the European Commission announced plans to ban thousands of carcinogenic and hormone-disrupting substances, including flame retardants, PFAS, and PVC. "If implemented," wrote the European Environmental Bureau, the continent's largest network of green groups, "the action will be the largest ever regulatory removal of authorised chemicals anywhere." Industry had raised a "storm of protest," the EEB reported, but EU member governments were solidly behind the initiative (though Italy opposed a PVC ban). The announcement came not long after the International Labour Organization released a report warning that more than one billion workers around the world are exposed to hazardous substances that cause "entirely preventable" deaths, illnesses, and injuries. Like the EPA, the ILO had compiled a "10 worst" list, which included asbestos, silica, solvents, nanomaterials, and a class of synthetic dyes, known as azo dyes, some of which degrade into aromatic amines—the ignominious family to which ortho-toluidine belongs. Azo dyes make a compelling case for regulating chemicals with similar characteristics in groups rather than individually. Any member of such a group would be presumed guilty unless its manufacturer could prove otherwise—the opposite of the system in place today.

The prospect of such a paradigm shift in politically riven America seems remote. And so, the safety of its workplaces will, for the foreseeable future, be a hit-or-miss proposition. Some are grungy throwbacks to the pre-OSHA era, others spotless. To understand how far the country has and hasn't come, it's instructive to watch some of the old government safety films that were screened in union halls and other settings. *The Hidden Hazards*, released by the US Public Health Service in 1963, was both prescient and naive. It explained how workers in the twentieth century were being afflicted by poisons that had bushwhacked their forebears: lead, silica, mercury. The scourge started with "the seemingly inexhaustible supply of immigrant labor" early in the century. "Many died," the narrator says. "Others returned to their homelands crippled and broken." Advances in technology in some ways had made things worse, as processes were installed without proper thought given to worker well-being. Fortunately, the field of industrial medicine had come into its own, state-employed industrial hygienists were "constantly on the watch" for dangerous substances, and most large companies, at least, had "outstanding records in occupational health." The atomic energy industry's record was "phenomenal."

This whopper was debunked decades later by the creation of a federal compensation program for nuclear workers who developed cancer after being exposed to plutonium and other radioactive materials. And it was clear by 1963 that most blue-collar industries put profit above employee health and the states were utterly incapable of regulating workplaces. The Public Health Service did get a few things right: there had been "steady and alarming" increases in chronic occupational diseases, it warned, and the death rate had "sharply climbed." Vigilance was in order, the film advised. Physicians needed to ask patients what they did for a living to flag these diseases early and prevent larger outbreaks. ("Ancient society," the health service pointed out, "had a simpler solution. It used slaves and criminals in the dangerous trades.")

Fifteen years after *The Hidden Hazards* was released, Eula Bingham's OSHA came out with *More Than a Paycheck*, narrated by John Wayne near the end of his life. It's somewhat surprising that Wayne, a lifelong conservative, would lend his name to a Carter administration production, though the actor had an independent streak and even attended Carter's inaugural ball. Wayne's voice sounds weak as he explains that one in four Americans would develop cancer—"those maverick cells"—and 3.5 million would be killed by the disease in the 1970s. He would succumb to stomach cancer the following year.

It was long thought that getting the disease was "a flip of the coin," Wayne says, but it had become clear that in most cases "cancer is caused by something we're exposed to." Some of these malignancies were being "sown in the workplace," and represented a risk "not as obvious as the giant ripsaw, the molten metal, the rigger's lofty perch." Exposures to chemicals or radiation "build up like a long-term bank account," Wayne says over the image of a man hacking during a breathing test. "The only difference is, this bank account pays off in a slow, lingering death, short-circuiting all the grand plans we made, wiping out those long-awaited golden years." The film includes an interview with physician and anti-asbestos crusader Irving Selikoff, who avers that "nature played a dirty trick on us" by attaching such a long latency period to cancer. Counting "one case at a time isn't meaningful," Selikoff says; it takes multiple cases to show a pattern, as happened when he studied death records for a cohort of union roofers. "It's only when we, in a sense, count the headstones that it suddenly occurs to us that something in the coke ovens or something in the dye vats are leading people to an early grave." The filmmaker cuts to a belching smokestack as tall as the Washington Monument at a copper smelter in Montana, and viewers hear that while the smelter provides five thousand jobs, it also dishes out lung cancer—perhaps from arsenic in the copper ore. Later they hear from Dr. Paul Kotin with the Johns-Manville Corporation, a maker of asbestos-containing insulation, roofing, and cement that suppressed studies linking the mineral to cancer and, smothered by its liabilities, would file for bankruptcy in 1982. Kotin shamelessly suggests that progress on occupational illness depends on "how well the worker cooperates" with good hygiene practices. He makes no mention of the employer's legal responsibility to keep the workplace safe. (And, in fact, court documents and testimony show that Manville had a policy well into the 1970s of not telling workers when their physical exams showed evidence of asbestos-related disease.) Wayne, our narrator, returns to float a concept that would have served companies like Goodyear well: instead of putting the onus on the worker, employers should either eliminate exposures or keep them to a minimum. "This," Wayne says, "means closing the system, sealing the process within closed vats, piping, tanks or what have you, keeping the dust, fumes or vapors out of workers' atmosphere." Closing the system isn't easy or cheap. But it's almost never as expensive as industry claims. And it's certainly not as costly as a factory full of sickness, and litigation derived from workers' untimely deaths.

Because regulation of toxics has failed so spectacularly, workers in the US are mostly left to their own devices. They are entitled by law to see material safety data sheets, which must lay out, in plain language, any dangerous properties of chemicals made or used at a particular workplace. In fact, these documents, prepared by manufacturers, are often unreliable and indecipherable. An analysis of 650 of the sheets in 2022 found that 30 percent included inaccurate warnings. One, for vinyl chloride, said the chemical could cause skin, eye, and respiratory irritation but failed to mention that it could cause cancer. Another, for benzene, had the same omission. All told, 15 percent of sheets for chemicals known to cause cancer did not identify them as carcinogens. DuPont's sheets had this failing while it was making ortho-toluidine, as did the placards it put on the rail tank cars that carried the chemical to Goodyear in Niagara Falls.

NIOSH director John Howard told me in early 2022 he was encouraged by recent EPA reviews of industrial chemicals produced in high volumes, such as the carcinogen trichloroethylene, or TCE. The agency was generating mountains of useful data, albeit slowly. If it banned or severely restricted a chemical that jeopardized workers as well as the public, as it finally did with methylene chloride in April 2023, OSHA could be spared from having to go through the tiresome standard-setting process. "We're in a hole now, so any way we can dig ourselves out with some of these high-volume products is a positive," Howard said. Indeed, even when a chemical has been studied to the nth degree and its victims can be counted and named, OSHA won't or can't respond. The standard for ortho-toluidine remains at 5 ppm, based on research last conducted in 1963—research that didn't consider the chemical's potential carcinogenicity. It took an improbable confluence of events—unrelenting pressure on the company from a strong union, an airtight health investigation by the federal government—even to uncover the bladder-cancer epidemic at Goodyear in Niagara Falls, let alone to address it. In 1983, about 20 percent of all workers in the United States belonged to a union; only about 10 percent belonged in 2022. If a cancer cluster like the one at Goodyear were percolating at a factory—or dozens of factories—at this moment, it might never be found.

The problem isn't one of science or technology. The tools to detect occupational disease are far superior to what was available a half-century ago. Since 2000, for example, Dr. Steven Markowitz, a physician and professor at the City University of New York who advised the OCAW and its successors, has used low-dose CT scans to detect lung cancer in a cohort of nearly fourteen thousand former and current US Department of Energy workers. He's found more than two hundred cases this way, three-quarters of which were in the early stages. But this is a government program. How can it be extended to the private sector? How can blue-collar workers at elevated risk of lung cancer be identified and enrolled in screening programs? There's no central exposure registry, no requirement that employers provide screenings after retirement or termination, no worker-education campaign. Few primary-care physicians bother to inquire about patients' occupations when examining them. The result? Stunning advances in medicine but an impoverished environment in which to apply them.

CHAPTER 25

"THIS STUFF
JUST DOESN'T GIVE UP"

I N 2022, HARRY WEIST and his in-laws, Ray and Dottie Kline, faced more tribulations, testing their already-fragile physical and emotional capacities. In February, Ray's bladder cancer returned for a third time. He'd been bleeding from his penis, and Rambarran, his urologist, discovered and removed a large tumor in the lining of the bladder. Ray went back on chemo. "This stuff just doesn't give up," Dottie told me by phone. Harry saw Rambarran the same month, took a prostate-specific antigen, or PSA, test and learned his number was high. The doctor told him he'd be dead of prostate cancer in a year, maybe a year and a half, if he didn't undergo radiation therapy. Harry, who'd resisted, agreed to return to New York for three months of treatment in the summer. He made a quick trip back in the late spring, undergoing a cysto on June 3. It came up clean but, as always, there had been complications. During his pre-operative exam at Mount St. Mary's Hospital three days earlier, his blood pressure was so high that the nurses had him sit in a wheelchair and took him to the emergency room. It was the first time he'd been in the ER since the night Diane died, and he felt flustered. Things got even more ticklish when the doctor who saw Harry was the same one who'd assured him it was safe for Diane to be discharged. "I said, 'Do you remember me?' The guy got a look of terror on his face." Harry's PSA level had dropped, but Rambarran said he would still need radiation. Ray, meanwhile, was unable to urinate, probably because of all the scar tissue in his urethra, and had to be fitted with a catheter and a bag. Ray went off the bag for a few days, still couldn't urinate, went back on it, went off, then back on. Harry was feeling pressure from his family to move from Las Vegas back to New York. He didn't rule it out. If something happened to Ray, he'd feel obligated to take care of Dottie.

Harry went home to Nevada for two weeks and was back in New York on June 22. When I talked to him almost a month later, he'd undergone ten days of radiation for his prostate cancer. "It's kicking my ass," he said. Before each treatment he had to drink water until his bladder was full and hold it while the radiologist blasted the tumor with a precisely targeted proton beam. The full bladder pushed the small intestine out of the way but caused Harry great pain. He was deeply tired much of the time and had trouble controlling his urine. The hair on his buttocks had fallen off. There were thirty-five more days of radiation to go.

One evening a few weeks earlier, Ray, Dottie, and Harry had been in the Klines' living room in Lewiston, watching a Western, when Dottie noticed Ray sleeping in an odd position in his recliner. "Sonny," she said, using her husband's nickname, "go get your pajamas on." Ray didn't respond. Harry got off the sofa and nudged his arm. Nothing. Harry lifted Ray from the chair but he remained limp and sunk back into the cushion. "My God, Bud," Dottie said in a panic. "I think he's dead." Harry called 911. While they waited for help, he kneeled next to Ray to see if his belly was rising and falling. It was—barely. Two Lewiston police officers arrived but were unable to rouse Ray. The paramedics who came afterward succeeded and asked Ray if he knew where he was. "I'm in Irvona, Pennsylvania, in my mom's house," he responded. They kept asking questions until Ray emerged from his trance.

Harry blamed the incident on exhaustion. At eighty-four, Ray was still delivering auto parts four days a week to help pay his and Dottie's medical bills. Harry had heard him swearing, uncharacteristically, when he got frustrated. Ray had been resistant to Harry's entreaties to quit the delivery job but had begun to soften, saying he'd consider scaling back to two days a week. "I'm worried about him," Harry told me. "I see him starting to lose his will."

On July 29, Ray Jr. hosted a gathering in Newfane to commemorate the first anniversary of Diane's death. Dottie and Ray Sr. were too ill to attend, and Harry figured this could be the last summer for both. The mood was lighter than it had been for Diane's memorial service ten months earlier. Holding a video camera, Ray Jr. went from one person to another—the three Weist children, their spouses, some former neighbors—and solicited stories about Diane. People remembered her as a worrier and a harsh critic of inattentive driving. When it was Harry's turn, he began reading from notes but choked up several times and had to stop, push up his glasses, and rub his eyes. He managed to get out a joke about Diane's obsession with shopping before closing with a sort of benediction: "Diane and I were always together, even when she took her last breath and went to be with Jesus in heaven. See you soon, Di. Love you so much."

Ray Jr. spoke the longest. When Diane graduated high school, her brother said, she weighed only ninety-eight pounds but could "knock the living crap" out of any boy in the neighborhood, including one who'd been teasing a mentally challenged girl Diane had befriended. Ray Jr. himself felt his sister's wrath—and surprisingly powerful punch—when he did something especially dense as a kid. But she taught him how to ride a bike and a motorcycle, shared his belief in a higher power, and remained a confidant the rest of her life.

The day after the gathering, Dottie, who suffered from congestive heart failure, was having trouble breathing and had to be taken by ambulance to Mount St. Mary's. It was the first of three trips to the emergency room. Twice Dottie was sent home, despite being violently ill. The third time Harry intervened. "We've seen this movie before," he told a nurse. "You sent Diane home and she died." Dottie was admitted, and for a time it appeared she wouldn't come out alive. "I'm gonna see Jesus, and I'm gonna see Diane," she told Harry at one point. She rallied, however, and was sent to a rehabilitation center, where she recovered from what turned out to be pneumonia. Ray, too, had seen progress. His catheter and bag had finally been removed, and he'd been taught how to catheterize himself if urine began to build up in his bladder. But he'd been shaken by Dottie's latest health scare and the idea of Harry going back to Nevada.

In late September, after finishing the last of his radiation treatments, Harry flew home to Las Vegas, which had lost its allure. "I sit on the couch and drink Crown [Royal]," he said. In New York, he'd heard from several women who seemed open to starting a relationship with him. The calls left him conflicted and a little flattered. He wasn't sure how his family would respond if he began seeing one or more of the women, but Diane had once insisted she didn't want him to lead a monastic life if she died. She'd be amused by his dilemma, Harry thought.

Nine months later, in June of 2023, Harry flew back to New York and went to Mount St. Mary's Hospital for what he felt certain would be a clean cystoscopy. It was not to be. Rambarran, the urologist, spotted and removed a tumor. It was Harry's fourth bout of bladder cancer, and he found the news deflating, to say the least. "I was confident," he said. "I'm like, 'Well, I had it in 2019 and 2020 and now it's been almost three years. I don't see me having a problem.'" Now he was back on a quarterly schedule of cystos, a procedure he always had done in New York because he didn't trust the doctors in Las Vegas. "Damn that Goodyear, man," Harry said.

POSTSCRIPT

GOODYEAR SHOULD BE remembered not only for its role, however muddled, in the Niagara Falls bladder-cancer tragedy but also for its presence among a group of PVC manufacturers that concealed the carcinogenic hazards of vinyl chloride from workers like Bill Smith, Harry Weist's colleague in Department 145. Goodyear, like the other companies, feigned surprise when the B.F. Goodrich angiosarcoma cluster in Louisville was revealed in January 1974. A Goodyear press release in March of that year stated that the company had undertaken an "extensive review" of employee medical records and found three angiosarcoma deaths. It was in the process of adopting new techniques to clean production equipment, adding ventilation, and developing more sophisticated devices to measure vinyl chloride in the air. It had reduced levels of the chemical in the plant to "a fraction" of what was then permitted by the government. It was, of course, telling only part of the story. The rest would come out much later.

In June of 2000, seventy-seven-year-old John Creech, the former Goodrich plant physician, gave only the second deposition of his life. Over four days of testimony in Louisville, Creech—tall and self-effacing, with a full head of white hair—would learn how his efforts to warn about the toxicity of vinyl chloride had been undermined without his knowledge. Despite his advanced age, Creech remembered the substance's odor from his time at Goodrich: "It smelled like a field of clover that has been cut down, a farmer making hay," the doctor said, responding to a question from Herschel Hobson, a cordial, balding lawyer from Beaumont, Texas, who was representing the family of a dead chemical worker. "It has kind of a sweetish—it's not really unpleasant—odor to it." The sweetness, Creech came to understand, masked the chemical's savage capacity to devour the liver and other organs and dissolve the bones in the hand. Creech, who grew up in the mine-scarred hills of Harlan County, Kentucky, said he always felt it was his obligation as a physician to be honest with his Goodrich patients about their prognoses and the origins of their illnesses—"just common, plain old human decency." This no doubt explains why, over those four days, he became increasingly dismayed and, in his gentlemanly way, enraged as Hobson showed him the most incriminating industry documents. Had Creech known about the secrecy agreement between the US manufacturers, including Goodrich, and their European counterparts regarding Cesare Maltoni's animal studies? He had not. Had he known that manufacturers in the early 1970s faced what one memo writer called "essentially unlimited liability" by selling vinyl chloride as a propellent in hairspray and other aerosol products, and that beauticians sometimes were hit with higher doses than PVC reactor cleaners? No again. (The Consumer Product Safety Commission banned this use of vinyl chloride in 1974.) On the last day of the deposition, self-doubt crept in. "I was rather proud of what I had done" at Goodrich, Creech, who died in 2018, said wanly. "All those big, successful accomplishments that you had back then have really shrunk down to the size of a chameleon. They are not dragons anymore. They are little chameleons."

The documents Creech reviewed had been acquired through discovery by Hobson and a lawyer in Lake Charles, Louisiana, named William Baggett Jr., who went by Billy.

Billy joined his father's law firm in 1982 and began working on asbestos cases, which he found unchallenging. He moved on to benzene and vinyl chloride cases; one of the latter would change the trajectory of his career and personal life. The case was *Ross v. Conoco*, brought in 1989 on behalf of Dan Ross, who was dying of brain cancer after working twenty-two years at a PVC plant in Lake Charles owned first by Conoco and then by a company called Condea Vista. Billy had been retained by Ross's steel-spined wife, Elaine, whom I met as a *Houston Chronicle* reporter in early 1998, eight years after Dan had died. Elaine was furious that her marriage had been cut short by what she believed to be corporate misconduct and wanted Billy to get to the root of it. The case, which was settled in 2001, would consume him for more than a decade, cost him a marriage, and send him to a rehabilitation center for six months.

The discovery process in *Ross* was fruitful in the extreme, yielding more than a million pages. The trove included damning communications from all the major PVC manufacturers and their primary trade group, the Chemical Manufacturers Association (now the American Chemistry Council). Industry studies of brain cancer among PVC workers showed evidence of data manipulation and suppression. I spent a year investigating the PVC industry, interviewing sick workers—and the families of dead ones—in Louisiana, Ohio, Mississippi, England, and Italy. When I was finished, I thought I'd never again see such an audacious example of corporate deception and disregard for worker health.

Then came ortho-toluidine. The bladder-cancer outbreak in Niagara Falls didn't make national headlines, as did vinyl chloride, but was a scandal hiding in plain sight. NIOSH faithfully tracked its progression and Steve Wodka represented its victims. DuPont blamed Goodyear; Goodyear blamed DuPont. Men like Hank Schiro suffered horribly and died. Families were broken.

When I met him in 2013, Ed Polka, the union safety man, habitually read the obituaries in the *Niagara Gazette* to see if anyone from Goodyear was among the fallen. "Holy Christ," he'd remark to his wife when he recognized a name. "You know who just passed away?" When I spoke with him nine years later, Polka, who retired in 2018, said he no longer subscribed to the newspaper and learned about such deaths only by word of mouth. He knew about Gary Casten but hadn't heard there was a seventy-eighth case of bladder cancer. His days were filled with grandchildren, household chores, fishing, hunting, and travel, and he gave Goodyear little thought. He believed the plant was "100 percent safer" than when he started in 1979, mainly because workers were wearing PPE. Would he feel comfortable working there today? He would. Did he believe exposure to ortho-toluidine could be eliminated? "I'd be a fool to say yes," Polka said.

Goodyear shows no sign of ending its use of the chemical. In 2021, the Niagara Falls plant received 2.56 million pounds from manufacturers in India and China. The same year, Goodyear reported net sales of $17.5 billion, a 42 percent increase over 2020 (the bump due mainly to its merger with Cooper Tire). It employed some seventy-two thousand people at fifty-seven sites in twenty-three countries. Its chairman, president, and CEO, Richard Kramer, received $21.4 million in total compensation. Ray Kline—octogenarian, bladder-cancer survivor, and loyal Goodyear employee for thirty-nine years—made $13 an hour delivering auto parts.

ACKNOWLEDGMENTS

THIS BOOK WOULD NOT HAVE BEEN POSSIBLE without the generosity of Steve Wodka, Harry and Diane Weist, and Ray and Dottie Kline. Over the course of a decade, Steve spent countless hours with me in person or by phone, explaining how a small chemical plant in western New York turned into a hothouse of disease. He opened up his voluminous legal and scientific files to me and guided me through the densest documents. The Weists and the Klines graciously shared their lives with me, painful parts and all. When Diane died suddenly in the summer of 2021, I considered Harry a dear friend, not the subject of a book, and heard the agony in his voice just hours after she passed.

There are many others to thank. Among them are Amy Caldwell and Susan Lumenello at Beacon Press; my agent, Esmond Harmsworth; the late, legendary Eula Bingham; lawyers William Baggett Jr., Amanda Hawes, Herschel Hobson, Tina Bradley, and Robert Wages; epidemiologist Peter Infante; physicians Robert Harrison and Steven Markowitz; environmental consultant Barry Castleman; John Howard, director of the National Institute for Occupational Safety and Health; retired Goodyear workers Ed Polka and Robert Dutton; helpful archivists at the Johnson, Nixon, and Reagan presidential libraries, and the Niagara Falls Public Library; longtime Niagara Falls resident and merchant Don King; *Niagara Gazette* reporter Rick Pfeiffer and retired reporter Don Glynn; and retired Service Employees International Union official Mark Catlin, whose collection of old worker-safety films was invaluable. Special thanks to the Whiting Foundation and the Fund for Investigative Journalism, which awarded me grants for this book; Harvard's Edmond & Lily Safra Center for Ethics for awarding me a fellowship; and my wife, Norma, son, Andrew, and daughter, Margaret, for enduring chest-high piles of documents, loud skirmishes with my computer (most of which I lost), and other vexations as I pulled together the manuscript.

REFERENCES

FROM 2013 TO 2023, I conducted dozens of interviews for this book with Harry Weist and Steve Wodka. Some were done in person—with Weist in Youngstown, Lewiston, and Niagara Falls, New York; and with Wodka in Buffalo, New York, and Little Silver, New Jersey. Many other interviews with Weist and Wodka were done by telephone or email. I met on several occasions with Weist's wife, Diane, and in-laws, Ray and Dottie Kline, in Youngstown and Lewiston, and spoke with them by phone.

Historical documents were accessed and photocopied at the National Archives in College Park, Maryland; the Lyndon B. Johnson, Richard Nixon, and Ronald Reagan presidential libraries; the Niagara Falls Public Library; and the University of Akron. Other documents were obtained from lawyers and scientists, including Steve Wodka, William Baggett Jr., Herschel Hobson, Tina Bradley, Amanda Hawes, Eula Bingham, Peter Infante, and Barry Castleman.

INTRODUCTION

Rodney Halford deposition, May 13, 1993.

National Institute for Occupational Safety and Health, Interim Report No. 1, Goodyear Tire and Rubber Company, Niagara Falls, New York, Dec. 1989.

John A. Zapp Jr. deposition, Mar. 2, 1987.

Tania Carreón et al., "Coronary Artery Disease and Cancer Mortality in a Cohort of Workers Exposed to Vinyl Chloride, Carbon Disulfide, Rotating Shift Work, and o-Toluidine at a Chemical Manufacturing Plant," *American Journal of Industrial Medicine* 57 (2014): 398–411.

Official transcript, hearings on H.R. 14816 before the Select Subcommittee on Labor of the Committee on Education and Labor, US House of

Representatives, Feb. 1, 20, 28, and 29 and Mar. 5, 6, 7, 8, 11, 12, and 14, 1968, Washington, DC.

Federal Register 79, no. 197 (Oct. 10, 2014).

AFL-CIO, "Death on the Job: The Toll of Neglect," Apr. 26, 2022, p. 1.

National Vital Statistics Reports, vol. 70, no. 9, US Centers for Disease Control and Prevention, July 26, 2021.

Jim Morris, Jamie Smith Hopkins, and Maryam Jameel, "Slow-Motion Tragedy for American Workers," Center for Public Integrity, June 29, 2015.

Office of the New York State Comptroller, 2012 Fiscal Profile, City of Niagara Falls.

"Niagara Falls," History.com, Mar. 4, 2010.

Mary Ellen Ellis, Mesothelioma.net (n.d.).

F. R. Holden, "What the Foundation Plant Surveys Are Disclosing," Seventh Annual Meeting of Members, Industrial Hygiene Foundation, Pittsburgh, PA, Nov. 10–11, 1942.

CHAPTER 1: NIAGARA FALLS BECOMES AN INDUSTRIAL LEVIATHAN

"Facts About Niagara Falls," Niagara Falls State Park (n.d.).

"Niagara Falls Historical Timeline," NYFalls.com (n.d.).

"Niagara Falls History," Niagara Falls State Park (n.d.).

Gerry Rising, "The Neutral Nation," *Buffalo Sunday News*, Sept. 6, 2009.

Robert L. McNamee, "Onaguiaahra," *Michigan Quarterly Review* (n.d.).

Buffalo History Gazette, Apr. 16, 2011.

Niagara Power Project exhibit, Lewiston, NY.

Raymond H. Arnot, "The Industries of Niagara Falls," *Popular Science Monthly* 73 (Oct. 1908).

"Invitation to Eastern Capitalists & Manufacturers," *National Intelligencer*, July 12, 1825.

George R. Shepard, "Industrial Niagara," *New York History* 18, no. 4 (Oct. 1937): 395–400.

John H. Lienhard, "The Dream Goes Wrong," *The Engines of Our Ingenuity* (n.d.).

Inez Whitaker Hunt, "Nikola Tesla," *Encyclopedia Britannica*, accessed Jan. 14, 2023.

"The Electro-Chemical Industries at Niagara Falls," *The Iron Age*, Apr. 3, 1902.

H. W. Buck, "Niagara Falls from the Economic Standpoint," *The Outlook*, May 19, 1906.

Historical Preservation Industrial Reconnaissance Survey, City of Niagara Falls, Nov. 2007.

"The Industries of Niagara Falls," Feb. 23, 1925, unidentified newspaper article.

Horace B. Brown, "Humming Turbines Chant Song of Returning Prosperity at Falls as Growing Power Demands Step Up Niagara Hudson Production," *New York Times*, (n.d.).

"Vital Defense Chemicals Being Produced at New Falls Factory," *Niagara Falls Gazette*, Dec. 5, 1941.

"DuPont Foremen Given Preview of Future Wonders," *Buffalo Courier-Express*, Feb. 25, 1945.

"Critical War Industries at Falls Menaced by Manpower Shortage," *Buffalo Courier-Express*, Feb. 18, 1945.

"Plastics Plant to Be Built Here by Goodyear Tire and Rubber Company," *Niagara Falls Gazette*, July 27, 1945.

"Welcome to Niagara" brochure, 1963, accessed at Niagara Falls Public Library.

"History of Goodyear," typewritten manuscript, accessed at University of Akron Goodyear archives.

North American Review CCVIIL (July 1865).

"Goodyear's Patent Extension," *Scientific American* (n.d.).

Transcript of Frank Seiberling speech (n.d.), accessed at University of Akron Goodyear archives.

Tom D. Crouch, "In the Museum: Dangerous Crossing," *Smithsonian Magazine*, Nov. 2010.

"Thirty Years of Goodyear," company publication, 1928, accessed at University of Akron Goodyear archives.

Associated Press, "Spreading Strike Shuts Tire Works," *New York Times*, Feb. 20, 1936.

Jack Sanders, "The American Mercury," Ridgefield Historical Society, Sept. 16, 2020.

"A Message to Goodyear Employees," *Akron Beacon Journal*, Aug. 22, 1945.

Karel Mulder and Marjolijn Knot, "PVC Plastic: A History of Systems Development and Entrenchment," *Technology in Society* 23, no. 2 (Apr. 2001): 265–86.

"Chemicals and Drugs: Industry Report," US Department of Commerce, Mar. 1949.

"The Application of Plastics to Peacetime Uses," transcript of R. P. Dinsmore speech, Cleveland Chamber of Commerce, Oct. 30, 1945, accessed at University of Akron Goodyear archives.

CHAPTER 2: RAY AND DOTTIE

"U.S. Synthetic Rubber Program: National Historic Chemical Landmark," American Chemical Society, Aug. 29, 1998, University of Akron.

Press release, Goodyear News Service, Sept. 10, 1958.

Press release, Goodyear News Service, June 7, 1960.

Deposition of John L. Creech, MD, June 12, 2000.

Memorandum from R. Emmet Kelly to A. G. Erdman, Jan. 7, 1966.

"MCA Medical Advisory Committee Meeting re PVC," Oct. 7, 1966.

Memorandum from George Roush to John L. Creech, Jan. 24, 1967.

Minutes of MCA Occupational Health Committee, Apr. 30, 1969.

Memorandum from George Ingle to W. E. Nessell (n.d.).

CHAPTER 3: AN AMERICAN "CASUALTY LIST"

Protecting the Health of Eighty Million Americans: A National Goal for Occupational Health, Special Report to the Surgeon General of the US Public Health Service, 2nd printing, Nov. 1966.

"Occupational Disease . . . the Silent Enemy." US Public Health Service (n.d.).

John D. Pomfret, "Johnson to Press Job Health Study," *New York Times*, May 24, 1966.

Memorandum from Ivan L. Bennett Jr. to Joseph Califano, Dec. 14, 1966.

Memorandum from Willard Wirtz to Joseph Califano, Dec. 5, 1967.

Official transcript, Hearings of the Select Subcommittee on Labor of the House Committee on Education and Labor, Feb. 1, 20, 28, and 29; Mar. 5, 6, 7, 8, 11, 12, and 14, 1968, pp. 12, 106–8, 188, 191, 204–6, 220, 232–35, 356, 359, 368, 387–88, 390, 739.

Official transcript, Hearings of the Senate Subcommittee on Labor of the Committee on Labor and Public Welfare, June 19, 1968, p. 350.

Congressional Record, May 24, 1968, p. 14918.

George C. Higgins, "The Yardstick: Attack by Chamber of Commerce," *Catholic Standard*, May 9, 1968.

CHAPTER 4: A NEW LAW, PROMPTLY ASSAILED

"Growers Spurn Negotiations on Poisons," *El Malcriado*, Jan. 15, 1969.

"What Are They Hiding?" *El Malcriado*, Feb. 1, 1969.

Statement of Dolores Huerta and Jerry Cohen, United Farm Workers, Nov. 21, 1969.

Transcript, "Hazards in the Industrial Environment," Oil, Chemical and Atomic Workers' Conference, Kenilworth, NJ, Mar. 29, 1969.

"James P. Mitchell Is Dead at 63; Eisenhower's Secretary of Labor," *New York Times*, Oct. 20, 1964.

Richard Nixon, "Special Message to the Congress on Occupational Safety and Health," transcript, Aug. 6, 1969.

Official transcript, Hearings of the House Select Subcommittee on Labor, Oct. 30 and Nov. 5, 6, 12, 13, and 18, 1969, p. 1181.

Official transcript, Hearing of the Senate Subcommittee on Labor, Mar. 7, 1970, pp. 821 and 822.

Senate Report No. 91–1282, Oct. 6, 1970, pp. 54–59.

Statement of Richard Nixon at signing of Occupational Safety and Health Act, Dec. 29, 1970.

Letter to Richard Nixon from John P. Hughes, Aug. 1, 1972, accessed at the Richard Nixon Presidential Library.

Letter from Rep. Dave Martin to George Guenther, Feb. 15, 1972, accessed at the Richard Nixon Presidential Library.

Letter from Sen. Clifford Hansen to James Hodgson, Mar. 30, 1972, accessed at the Richard Nixon Presidential Library.

Letter from O. William Habel to Richard Nixon, Oct. 11, 1971, accessed at the Richard Nixon Presidential Library.

"Federal Safety Act Will Cost Public, Industry Millions," *Grand Rapids Press*, June 9, 1971.

Jim Morris, "After 44 Years, Halting Progress on Workplace Disease," Center for Public Integrity, July 6, 2015.

Occupational Safety and Health Administration citation issued to Allied Chemical Corporation, May 28, 1971.

Les Leopold, *The Man Who Hated Work and Loved Labor: The Life and Times of Tony Mazzocchi* (White River Junction, VT: Chelsea Green, 2007), p. 287.

CHAPTER 5: TYLER'S ASBESTOS DISASTER

Industrial Hygiene Foundation of America, *Evaluation of the Asbestos Dust Hazard in Tyler, Texas Plant*, report to Pittsburgh Corning Corporation, July 8 and Aug. 6–7, 1963.

J. T. Destefano, *Industrial Hygiene Survey, Pittsburgh Corning, Tyler, Texas*, report issued Sept. 11, 1967.

Letter from William M. Johnson, MD, to Steve Wodka, Sept. 24, 1971.

Letter from Jeremiah R. Lynch to J. W. McMillan, Mar. 27, 1968.

Letter from Clinton V. Oster to C. E. Van Horne, Jan. 27, 1970.

Memorandum from William M. Johnson, MD, to Edward J. Fairchild, Aug. 17, 1971.

Homer Bigart, "Lung-Disease Problem, Traced to Beryllium Refinery, Plagues Hazleton, Pa.," *New York Times*, Oct. 29, 1972.

"NIOSH Survey, Pittsburgh Corning Corporation, Tyler, Texas," Oct. 26–29, 1971.

Letter from William Farkos to J. B. Stokes, Nov. 1, 1971.

OSHA inspection report, Dec. 14, 1971.

OSHA citation issued to Pittsburgh Corning, Dec. 16, 1971.

OSHA citation issued to Pittsburgh Corning, Jan. 13, 1972.

NIOSH report on burlap bags, prepared by Roy M. Fleming, Jan. 14, 1972.

Telegram from A. F. Grospiron to George Guenther, Feb. 8, 1972.

Letter from George Guenther to A. F. Grospiron, Feb. 16, 1972.

Letter from William M. Johnson, MD, to James E. Peavy, MD, Nov. 16, 1972.

Paul Brodeur, "Casualties of the Workplace," pt. 2, *New Yorker*, Nov. 5, 1973.

Letter from John R. Quarles Jr. to A. F. Grospiron, Feb. 18, 1972.

NIOSH report by John M. Dement, Philip J. Bierbaum, and Ralph D. Zumwalde, Sept. 6, 1973.

Letter from William M. Johnson, MD, to the *Journal of Occupational Medicine* 16, no. 10 (Oct. 1974).

"Court Clears Payment of $20 Million in Suit by Asbestos Workers," *Wall Street Journal*, Feb. 10, 1978.

Interview with Jeffrey L. Levin, MD.

Jeffrey L. Levin et al., "Tyler Asbestos Workers: Mortality Experience in a Cohort Exposed to Amosite," *Occupational and Environmental Medicine* 55 (1998): 155–60.

CHAPTER 6: VINYL

Center for Disease Control, *Morbidity and Mortality Weekly Report* 23, no. 6 (Feb. 15, 1974): 49–50.

Jane E. Brody, "Plastics Workers Screened for Ill Effects of Vinyl Chloride," *New York Times*, Mar. 13, 1974.

Society of the Plastics Industry, Inc. v. Occupational Safety and Health Administration (Jan. 31, 1975), accessed in the *Environmental Law Reporter*.

Peter F. Infante, "Oncogenic and Mutagenic Risks in Communities with Polyvinyl Chloride Production Facilities," *Annals of the New York Academy of Sciences* 271 (May 1976): 49–57.

P. F. Infante, J. K. Wagoner, and R. J. Waxweiler, "Carcinogenic, Mutagenic and Teratogenic Risks Associated with Vinyl Chloride," *Mutation Research* 41 (Nov. 1, 1976): 131–41.

Letter from J. William Flynt, MD, to Joseph K. Wagoner, Oct. 28, 1975.

Memorandum from Joseph K. Wagoner and Peter F. Infante to Jack Finklea, Nov. 5, 1975.

Jane E. Brody, "2D Region Studied for Birth Defects," *New York Times*, Mar. 5, 1976.

Interview with Peter Infante.

F. A. Patty, W. P. Yant, and C. P. Waite, "Acute Response of Guinea Pigs to Vapors of Some New Commercial Organic Compounds: Vinyl Chloride," *Public Health Reports* 45, no. 34 (Aug. 22, 1930): 1963–71.

S. L. Tribukh et al., "Working Conditions and Measures for Their Improvement in Production and Use of Vinyl Chloride Plastics," *Gigiena i Sanitariya* 10 (1949): 38–44, cited in "Environmental Chemicals: Human and Animal Health," Proceedings of 4th Annual Conference, Colorado State University, July 7–11, 1975.

T. R. Torkelson, F. Oyen, and V. K. Rowe, "The Toxicity of Vinyl Chloride as Determined by Repeated Exposure of Laboratory Animals," *American Industrial Hygiene Association Journal* 22, no. 5 (1961): 354–61.

Memorandum from Henry F. Smyth Jr. to T. W. Nale, Nov. 24, 1959.

Letter from V. K. Rowe to W. E. McCormick, May 12, 1959.

Memorandum from W. Mayo Smith to Richard Fleming, June 6, 1974.

P. L. Viola, A. Bigotti, and A. Caputo, "Oncogenic Response of Rat Skin, Lungs, and Bones to Vinyl Chloride," *Cancer Research* 31, no. 5 (May 1971): 516–22.

Letter from George Roush to Richard Henderson, June 24, 1970.

Report of the CMA Technical Task Group on Vinyl Chloride Research, Nov. 14, 1972.

Deposition of Marcus M. Key, MD, Sept. 19, 1995.

Jennifer Beth Sass, Barry Castleman, and David Wallinga, "Vinyl Chloride: A Case Study of Data Suppression and Misrepresentation," *Environmental Health Perspectives* 113, no. 7 (July 2005).

Deposition of Holly Smith, June 29, 1995.

CHAPTER 7: HARRY BREAKS FREE

Interviews with Harry Weist and Steve Wodka.

David Burnham, "Death of Plutonium Worker Questioned by Union Official," *New York Times*, Nov. 19, 1974.

The Reasoner Report, transcript, ABC News, Mar. 1, 1975.

CHAPTER 8: EULA

Interview with Eula Bingham.

"Leukemia Killed 7 at Goodyear Who Worked with Same Chemical," *Akron Beacon Journal*, Apr. 18, 1976.

"11 Deaths Now Total in Pliofilm," *Akron Beacon Journal*, May 13, 1976.

Letter from Peter Bommarito to William Usery, April 23, 1976.

Post-Hearing Brief of United Steelworkers of America on Standard for Coke Oven Emissions, June 16, 1976.

Statement by Ralph Nader and Sidney Wolfe, Birmingham, England, May 19, 1976.

Letter from George Meany to William Usery, Mar. 5, 1976.

Letter from George Meany to William Usery, Sept. 27, 1976.

"Cancer and the Environment: A Scientific Perspective," Samuel S. Epstein, MD, Feb. 1976.

"Why OSHA Should Be Dissolved," *Factory*, Aug. 1976.

David Burnham, "Agency Assailed by Ford Defers New Safety Rules till after Election," *New York Times*, Mar. 4, 1976.

"Status Report on OSHA," Morton Corn, Jan. 12, 1977.

Proceedings, Conference on Women and the Workplace, Washington, DC, June 17–19, 1976.

Katharine Q. Seelye, "Eula Bingham, Champion of Worker Safety, Dies at 90," *New York Times*, June 23, 2020.

Eula Bingham press conference, Apr. 29, 1977, accessed via YouTube.

Whitt Flora, "Dr. Bingham Is Blowing Fresh Air into OSHA," *Cincinnati Post & Times-Star*, Apr. 2, 1977.

Industrial Union Department, AFL-CIO, v. American Petroleum Institute, decided July 2, 1980, accessed via Justia.

The Today Show, transcript, May 20, 1977.

Sir Thomas Oliver, MD, "Lead Poisoning and the Race," lecture, Eugenics Education Society, London, May 4, 1911.

Philip Shabecoff, "The Woman Who Turned OSHA Around," *New York Times*, Feb. 19, 1978.

"Potential Health Risks to DOD Firing-Range Personnel from Recurrent Lead Exposure," National Research Council, Dec. 3, 2012.

Transcript of testimony of the United Steelworkers of America, July 11, 1978.

Telegram from A. F. Grospiron to Ray Marshall, July 14, 1977.

Carol Kleiman, "Sterilized or Fired? The Choice Was Theirs," *Chicago Tribune*, Jan. 28, 1979.

Letter from H. Christine Whiteman to Steve Wodka, Feb. 19, 1979.

United States Court of Appeals for the District of Columbia Circuit, No. 81-1687, Oil, Chemical and Atomic Workers International Union and Local 3–499, Oil, Chemical and Atomic Workers v. American Cyanamid Company (1982).

A. Dahleen Glanton, "A Woman's Place . . . : Plant's Policy to Protect Unborn Clashes with Job Bias Rules," *Los Angeles Times*, May 15, 1988.

"Cotton Dust: Worker Health Alert," Occupational Safety and Health Administration, 1980.

CHAPTER 9: A BLUE-COLLAR SOCIAL CLUB

Norma Higgs, "A Look at Life in the Falls in the '50s," *Niagara Gazette*, May 25, 2015.

Deposition of Henry T. Schiro, Nov. 4, 1985.

Memorandum from Art Zimmerman to Bruce Bendow, Nov. 9, 1960.

Press release, Goodyear News Bureau, Feb. 12, 1962, accessed at the University of Akron Goodyear archives.

Photo and caption, Goodyear News Bureau, 1957 (n.d.), accessed at the University of Akron Goodyear archives.

Press release, Goodyear News Bureau, Aug. 14, 1967, accessed at the University of Akron Goodyear archives.

Press release, Goodyear News Bureau, 1976 (n.d.), accessed at the University of Akron Goodyear archives.

Press release, Goodyear News Bureau, 1979 (n.d.), accessed at the University of Akron Goodyear archives.

"Schoellkopf Power Plant Ruins Site," Niagara Falls State Park.

Sidney H. Schanberg, "Slump a Worry at Niagara Falls," *New York Times*, May 26, 1963.

"A $26 Million Facelift," *Globe and Mail*, May 5, 1967.

"AP Falls Story Due," *Niagara Gazette*, Jan. 8, 1974.

"Falls Renewal Unit Grants an Option for Rotating Hotel," *Buffalo Courier-Express*, Feb. 16, 1972.

Rita Reif, "Blighted Niagara Falls Works to Rebuild Itself," *New York Times*, Oct. 1, 1973.

Mike Brown, "Red Tape Stalls Dump Solution," *Niagara Gazette*, Feb. 5, 1978.

Eckhardt C. Beck, "The Love Canal Tragedy," *EPA Journal*, Jan. 1979.

Dr. Jordan Kleiman, "Love Canal: A Brief History," State University of New York Geneseo (n.d.).

Joanne Omang, "Love Canal Families to Be Relocated," *Washington Post,* May 22, 1980.

CHAPTER 10: DUPONT AND DOMINIC

"Toluene," American Chemical Society, Mar. 18, 2019.

"History," Dupont.com (n.d.).

Holger Georg Dietrich and Klaus Golka, "Bladder Tumors and Aromatic Amines—Historical Milestones from Ludwig Rehn to Wilhelm Hueper," *Frontiers in Bioscience* 4, no. 1 (Jan. 2012): 279–88.

"Cancer of the Bladder among Workers in Aniline Factories," International Labour Office, Jan. 1921.

"A Challenge That Was Accepted," *DuPont Magazine,* Sept. 1935.

Silas Dent, "Tetraethyl Lead Fatal to Makers," *New York Times,* June 22, 1925.

David Michaels, "When Science Isn't Enough: Wilhelm Hueper, Robert A. M. Case, and the Limits of Scientific Evidence in Preventing Occupational Bladder Cancer," *International Journal of Occupational and Environmental Health* 1 (1995): 278–88.

"The Carcinogenic Agent—Chemistry and Industrial Aspects," transcript of presentation by G. H. Gehrmann, 1934 (n.d.).

"Protecting the Public Health," DuPont publication (n.d.).

Wilhelm C. Hueper, "Adventures of a Physician in Occupational Cancer: A Medical Cassandra's Tale," unpublished manuscript, 1976.

Memorandum from Robert J. Weiss, MD, to William E. Fayerweather and M. Elizabeth Karns, Oct. 25, 1991.

David Michaels, "Ortho-Toluidine and Human Bladder Cancer," expert report to Steve Wodka, Mar. 25, 2008.

Deposition of John A. Zapp Jr., Mar. 2, 1987.

James Huff, "Benzene-Induced Cancers: Abridged History and Occupational Health Impact," *International Journal of Occupational and Environmental Health* 13, no. 2 (2007): 213–21.

Memorandum from Nancy J. Hunt to E. F. Schultz, June 12, 1974.

Memorandum from B. C. McKusick to S. N. Boyd, Dec. 23, 1974.

Memorandum from E. F. Schultz to DuPont customers, Jan. 12, 1977.

M. I. Khlebnikova et al., "Problems of Industrial Hygiene and Health Status of Workers Employed in the Production of o-Toluidine," 1975 English translation of 1970 study by Soviet Institute of Industrial Hygiene and Diseases.

Memorandum from S. N. Boyd Jr. to E. F. Schultz, Feb. 21, 1977.

Kevin W. Hanley et al., "Exposure to o-Toluidine, Aniline, and Nitrobenzene in a Rubber Chemical Manufacturing Plant: A Retrospective Exposure Assessment Update," *Journal of Occupational and Environmental Hygiene* 9 (Aug. 2012): 478–90.

Carl B. Kaufmann, "A 5-Part Quiz on Corporate Ethics," *Washington Post*, July 1, 1979.
Barry Castleman, "DuPont's Record in Business Ethics: Another View," *Washington Post*, July 15, 1979.

CHAPTER 11: CANCER ERUPTS AT GOODYEAR
Interview with Christine Oliver.
Christine Oliver report to OCAW Local 8–277, Mar. 2, 1979.
Industrial Hygiene Survey, Department 245, Goodyear Tire and Rubber Company, Niagara Falls, Robert F. Gempel, Feb. 16, 1980.
L. Christine Oliver and R. P. Weber, "Chest Pain in Rubber Chemical Workers Exposed to Carbon Disulphide and Methaemoglobin Formers," *British Journal of Industrial Medicine* 41 (1984): 296–304.
Letter from Rodney Halford to James Pearson, Mar. 25, 1981.
Letter from James Pearson to James Ward, Apr. 10, 1981.
Memorandum from Christine Oliver to Jim Ward, July 16, 1981.
Deposition of Rodney Halford, May 13, 1993.
Deposition of Clifford A. Johnson, MD, Apr. 9, 1991.
Deposition of Henry T. Schiro, Nov. 4, 1985.
Interview with Bob Bailey.

CHAPTER 12: REAGAN
Interview with Steve Wodka.
Ronald Reagan, transcript of Election Eve address, Nov. 3, 1980.
Executive Order 12291, Feb. 17, 1981.
"Thorne Auchter Administration, 1981–1984: 'Oh, What a (Regulatory) Relief,'" US Department of Labor (n.d.).
Statement of Thorne G. Auchter to Senate Subcommittee on Investigations and General Oversight, Sept. 23, 1981.
Felicity Barringer, "Judge Orders OSHA to Toughen Standard on Exposure to Gas," *Washington Post*, Jan. 7, 1983.
Occupational Safety and Health Committee report to Chemical Manufacturers Association Board of Directors, Jan. 11, 1982.
Letter from Bailus Walker to Peter Infante, June 29, 1981.
Letter from US Rep. Albert Gore Jr. to Thorne Auchter, July 1, 1981.
Morton Mintz, "Sparks Fly over Attempt to Fire OSHA Expert," *Washington Post*, July 17, 1981.
Morton Mintz, "OSHA Backs Off Trying to Fire Cancer Specialist," *Washington Post*, Aug. 10, 1981.
IARC Monograph 29, formaldehyde, 1982.
Thorne Auchter press conference, Mar. 27, 1981, accessed via YouTube.
Memorandum from Bob Bonitati to Elizabeth Dole and Red Cavaney, Apr. 27, 1981.

Cristine Russell, "Nader Group Study Calls OSHA Too Lax," *Washington Post*, Nov. 9, 1982.
Letter from Ralph Nader to Ronald Reagan, Nov. 22, 1983.
Deposition of Richard Kevin Sullivan, Apr. 6, 1989.
Deposition of Henry T. Schiro, Nov. 4, 1985.

CHAPTER 13: HARRY MOVES UP
Deposition of Harry Weist, Jan. 31, 2006.
Letter from Steve Wodka to Robert Wages, Nov. 12, 1987.
Letter from Sylvia Krekel to William Halperin, Feb. 4, 1988.
National Institute for Occupational Safety and Health, Interim Report No. 1, Goodyear Tire and Rubber Company, Niagara Falls, New York, Dec. 1989.
Elizabeth Ward et al., "Excess Number of Bladder Cancers in Workers Exposed to Ortho-Toluidine and Aniline," *Journal of the National Cancer Institute* 83, no. 7 (Apr. 1991): 461–524.
Adrian L. Linch, "Biological Monitoring for Industrial Exposure to Cyanogenic Aromatic Nitro and Amino Compounds," *American Industrial Hygiene Association Journal* 35, no. 7 (1974): 426–32.
Deposition of John A. Zapp Jr., Mar. 2, 1987.
Deposition of Joseph L. Holtshouser, Aug. 5, 2010.
Letter from George Simmons to Gary Chaffee, June 21, 1991.

CHAPTER 14: WHAT IS BLADDER CANCER?
Khurshid A. Guru, "A Patient's Guide to Bladder Cancer," Roswell Park Comprehensive Cancer Center, July 2017.
"What Is Bladder Cancer?," American Cancer Society (n.d.).
Deposition of Joseph T. Nicastro, Nov. 10, 2008.
Deposition of Joseph T. Nicastro, Jan. 16, 2009.
Deposition of Joseph T. Nicastro, Feb. 18, 2010.
Interview with Pamela Nicastro.

CHAPTER 15: THE GOODYEAR EPIDEMIC SPREADS
Lawrence J. Fine, MD, and Elizabeth Ward, letter to Goodyear workers, June 11, 1990.
Interview with Ed Polka.
Interview with Philip Aliotta.
Deposition of Philip Aliotta, MD, May 20, 1999.
Elizabeth Ward et al., "Excess Number of Bladder Cancers in Workers Exposed to Ortho-Toluidine and Aniline," *Journal of the National Cancer Institute* 83, no. 7 (Apr. 1991): 461–524.
Memorandum from Joseph T. Holtshouser to A. H. Olzinger, Apr. 9, 1992.
Letter from Donald J. Sherman, MD, to Elizabeth Ward, May 23, 1996.
Interview with Robert Dutton.

Deposition of Rodney Halford, May 13, 1993.
Deposition of Richard E. Prato, Oct. 15, 1996.
Deposition of Dorothy J. Kowalski, Jan. 3, 1995.
Deposition of Louis Kowalski Jr., Mar. 7, 1995.

CHAPTER 16: RAY AND HARRY GET BAD NEWS
"Goodyear Cutting 166 Jobs at Niagara Falls Plant," *Buffalo News*, June 28, 1996.
Deposition of Raymond Kline, July 14–16, 1998.
"Uncertain Future Faced by Workers at Goodyear Plant," *Buffalo News*, Mar. 21, 2003.
Ken Ward Jr., "Justices Unseal DuPont C8 Documents. Memos Detail Company Lawyers' View on Vulnerability to Pollution Suit," *Charleston Gazette*, May 7, 2004.
Deposition of Joseph Nicastro, Jan. 16, 2009.
Deposition of Pamela Nicastro, Jan. 16, 2009.

CHAPTER 17: WODKA FORTIFIES HIS CASE AGAINST DUPONT
Deposition of James A. Medaris, Apr. 3, 2012.
Deposition of Barbara Dawson, Apr. 7, 2016.
Deposition of Thomas Nelson, Jan. 14, 2016.
New York State Workers' Compensation Board, amended administrative decision in regard to Guy Mort, Sept. 23, 2019.

CHAPTER 18: CHEMICALS ARE OUT OF CONTROL
"Mismanaging Chemical Risks: EPA's Failure to Protect Workers," Hearing before the Subcommittee on Environment and Climate Change of the Committee on Energy and Commerce, US House of Representatives, Mar. 13, 2019.
W. C. Hueper, *Occupational Tumors and Allied Diseases* (Springfield, IL: Charles C. Thomas, 1942).
Henry F. Smyth, "Solving the Problem of the Toxicity of New Chemicals in Industry," *West Virginia Medical Journal* 7, no. 42 (1946).
Ralph Landau and Ashish Arora, "The Chemical Industry: From the 1850s until Today: Growing by Restructuring and Adapting to Changing Environments," *Business Economics* 34, no. 4 (Oct. 1999): 7–15.
"API Toxicological Review: Benzene," Sept. 1948.
Memorandum from C. H. Hine, MD, to K. R. Edlund, Apr. 28, 1950.
W. C. Hueper, MD, "Cancer of the Urinary Bladder in Workers of Chemical Dye Factories and Dyeing Establishments: A Review," *Journal of Industrial Hygiene* 16, no. 5 (Sept. 1934).
M. W. Goldblatt, "Vesical Tumours Induced by Chemical Compounds," *British Journal of Industrial Medicine* 6, no. 2 (1949): 65–81.

Alan Derickson, "Inventing the Right to Know: Herbert Abrams's Efforts to Democratize Access to Workplace Health Hazard Information in the 1950s," *American Journal of Public Health* 106, no. 2 (Feb. 2016).

Rachel Carson, *Silent Spring* (Boston: Houghton Mifflin, 1962).

"The U.S. Federal Government Responds," Environment and Society Portal (n.d.).

Council on Environmental Quality report, Apr. 1971.

"Oral History Interview of the Toxic Substances Control Act from the Perspective of J. Clarence Davies," Science History Institute, Oct. 30, 2009.

"Off the Books II: More Secret Chemicals," Environmental Working Group (n.d.).

Richard A. Denison, "A Primer on the New Toxic Substances Control Act (TSCA) and What Led to It," Environmental Defense Fund, Apr. 2017.

Barry Commoner, "The Promise and Perils of Petrochemicals," *New York Times*, Sept. 25, 1977.

Sandy Smith, "The First 10: The Chemicals EPA Will Review under New TSCA Legislation," *EHS Today*, Nov. 29, 2016.

"How to Access the TSCA Inventory," US Environmental Protection Agency, Feb. 2022.

Sheldon Krimsky, "The Unsteady State and Inertia of Chemical Regulation under the US Toxic Substances Control Act," *PLOS Biology* 15, no. 12 (Dec. 2017).

Press release, American Chemistry Council, Apr. 14, 2022.

Summary of American Chemistry Council lobbying, Opensecrets.org, accessed July 6, 2023.

CHAPTER 19: OLD SCOURGES REVISITED

"The Third Wave of Asbestos Disease: Exposure to Asbestos in Place," bound volume published by the New York Academy of Sciences, 1991.

Interview with Kris Penny.

"Investigation Relating to Health Conditions of Workers Employed in the Construction and Maintenance of Public Utilities," US House of Representatives Committee on Labor, Jan. 16, 1936, transcript accessed via Google Books.

Stop Silicosis, US Department of Labor (1938), accessed via YouTube.

Interview with Fernando Salmeron.

Interviews with Rodrigo and Dora Alicia Martinez.

OSHA inspection of Stone Etc., Hayward, CA, opened Jan. 28, 2019.

OSHA inspection of Stone Etc., Gardena, CA, opened Feb. 5, 2019.

"Severe Silicosis in Engineered Stone Fabrication Workers—California, Colorado, Texas, and Washington, 2017–2019," *Morbidity and Mortality Weekly Report*, Centers for Disease Control and Prevention, Sept. 27, 2019.

"A Public Health Pioneer," *Think*, online magazine for Case Western Reserve University (n.d.).

Interview with Robert Harrison.

Robert Harrison et al., *Merck Is Not a Candy Factory*, medical electives report to Oil, Chemical and Atomic Workers Union, Aug. 16, 1976.

Interviews with Gustavo Reyes Gonzalez and Juan Gonzalez Morin.
Interviews with Jane Fazio and Nader Kamangar.

CHAPTER 20: KIDS
Interview with Yvette and Mark Flores.
Interview with Amanda Hawes.
Interview with LeeAnn Severson.
Washington State Department of Social and Health Services, assessment of Darryl K. Severson, Aug. 4, 2021.
Interview with Cynthia Bearer.
Alice Hamilton, MD, *Industrial Poisons in the United States* (New York: The Macmillan Company, 1925).
Sir Thomas Oliver, MD, "Lead Poisoning and the Race," lecture delivered to the Eugenics Education Society, London, May 4, 1911.
Dorothy E. Donley, "Toxic Encephalopathy and Volatile Solvents in Industry: Reports of a Case," *Journal of Industrial Hygiene and Toxicology* 18 (Oct. 1936).
Leonard Greenburg et al., "Health Hazards in the Manufacture of 'Fused Collars,'" *Journal of Industrial Hygiene and Toxicology* 20, no. 2 (1937).
"Medicine and the War," *Journal of the American Medical Association* (Sept. 5, 1942).
"Health Hazards from Industrial Solvents," *War Department Technical Bulletin* (Apr. 27, 1944).
Anna M. Baetjer, *Women in Industry: Their Health and Efficiency* (National Research Council, 1946).
W. MacLeod Ross, "Environmental Problems in the Production of Printed Circuits," *Annals of Occupational Hygiene* 15 (1972): 141–51.
Proceedings, Conference on Women and the Workplace, Washington, DC, June 17–19, 1976.
Health Hazard Evaluation Project No. HHE 79–66, Interim Report #2, Signetics Corporation, National Institute for Occupational Safety and Health, Jan. 30, 1980.
H. Pastides et al., "Spontaneous Abortion and General Illness Symptoms among Semiconductor Manufacturers," *Journal of Occupational Medicine* 30, no. 7 (1988): 543–51.
S. H. Swan et al., "Historical Cohort Study of Spontaneous Abortion among Fabrication Workers in the Semiconductor Health Study: Agent-Level Analysis," *American Journal of Industrial Medicine* 28, no. 6 (1995): 751–69.
Marcus LeDeaux et al., v. Motorola Incorporated and/or Motorola Solutions, plaintiffs' memorandum in support of punitive damages, 2018 IL App (1st) 161345, filed Dec. 20, 2018.
Marcus LeDeaux et al., v. Motorola Incorporated and/or Motorola Solutions, defendant's response to plaintiffs' memorandum in support of punitive damages, Circuit Court of Cook County, Illinois, Case No. 10-L-8503, filed Jan. 28, 2019.

"Motorola, Inc.," *Britannica Online Encyclopedia*, accessed June 4, 2022.
Deposition of Robert G. Numkena, June 10, 2014.
Deposition of Erma Acosta, Mar. 1, 2013.

CHAPTER 21: RAY AND HARRY IN RETIREMENT
Interviews with Harry Weist et al.

CHAPTER 22: WODKA FINISHES THE FIGHT
Report and Recommendation of the US Magistrate Judge, Hon. H. Kenneth
    Schroeder, filed July 19, 2021, in the case of Douglas J. and Suzanne M.
    Moss v. E. I. duPont de Nemours and Co.
Interview with Jim and Debbie Sarkees.
Interview with Gary Casten.
Deposition of Gary Casten, Apr. 22, 2021.
Complaint filed with Environmental Protection Agency by Steve Wodka, Jan.
    4, 2021.
Press release by Environmental Protection Agency, Dec. 15, 2005.
Robert M. Park, Tania Carreon, and Kevin Hanley, "Risk Assessment for
    O-Toluidine and Bladder Cancer Incidence," *American Journal of Indus-
    trial Medicine* 64 (2021): 758–70.
Letter from Mark Kitchen to Joseph White and Steve Wodka, July 13, 2021.
Memorandum from Gordon R. Stratton and H. Tim Frazer to Vicki Fulimeni,
    Oct. 14, 2021.
Letter from Joseph White and Steve Wodka to Mark Kitchen, Oct. 19, 2021.
Email from Steve Sallman, United Steelworkers, to Jim Morris, Mar. 30, 2022.
Interview with Robert Wages.

CHAPTER 23: WODKA'S CRUCIBLE
Operative record on Edward Babut, Niagara Falls Memorial Medical Center,
    June 14, 1973.
State of New York Workmen's Compensation Board Notice of Decision, Ed-
    ward Babut, claimant, Nov. 20, 1973.
Semi-Annual Report—Spring 2022 Bladder Cancer Screening Program, Ever-
    side Health, July 22, 2022.
Email from Daniel A. Turney, director of public affairs, DuPont, to Jim Mor-
    ris, Aug. 3, 2022.
Email from Connie Deibel, communications manager, global operations,
    Goodyear, to Jim Morris, Aug. 4, 2022.

CHAPTER 24: WORKERS ARE (MOSTLY) ON THEIR OWN
W. J. Nicholson and P. J. Landrigan, "Quantitative Assessment of Lives Lost
    due to Delay in the Regulation of Occupational Exposure to Benzene,"
    Environmental Health Perspectives 82 (July 1989): 185–88.

Final rule: "Air Contaminants, Occupational Safety and Health Administration," *Federal Register* 54 (Jan. 1989): 2332–983.

Memorandum from Roger A. Clark to OSHA directorate heads and regional administrators, Aug. 5, 1993.

"OSHA Listens: Occupational Safety and Health Administration Stakeholder Meeting," *Federal Register* 75 (Jan. 2010): 2890–91.

OSHA national news release, Oct. 24, 2013.

Final rule: "Occupational Exposure to Respirable Crystalline Silica," *Federal Register* 81, no. 58 (Mar. 2016).

United Steelworkers news release, Jan. 9, 2017.

AFL-CIO, "Death on the Job: The Toll of Neglect," Apr. 26, 2022.

"2021 Proposed ACGIH Benzene TLV," American Conference of Governmental Industrial Hygienists.

"Committee for Risk Assessment: Opinion on Scientific Evaluation of Occupational Exposure Limits for Benzene," European Chemicals Agency, Mar. 9, 2018.

Interviews with Chuck Geraci and Jay Vietas.

Press release, European Environmental Bureau, Apr. 25, 2022.

"Exposure to Hazardous Chemicals at Work and Resulting Health Impacts: A Global Review," International Labour Organization, May 7, 2021.

The Hidden Hazards, US Public Health Service (1963), accessed via YouTube.

More Than a Paycheck, Occupational Safety and Health Administration (1978), accessed via YouTube.

"Obstructing the Right to Know," BlueGreen Alliance and Clearya (2022).

Interview with John Howard.

"Union Membership Fell to Record Low in 2022, Bureau of Labor Statistics Says," CNBC.com, Jan. 19, 2023.

Interview with Steven Markowitz.

CHAPTER 25: "THIS STUFF JUST DOESN'T GIVE UP"

Interviews with Harry Weist.

Video of Diane Weist remembrance, July 29, 2022, provided by Ray Kline Jr.

Interview with Rick Pfeiffer.

POSTSCRIPT

Deposition of John L. Creech, MD, June 12, 2000.

Interviews with Ed Polka.

Trade data compiled by Material Research, L3C.

Goodyear news release, Feb. 11, 2022.

"Richard Kramer," Economic Research Institute website (n.d.).

INDEX